DESTINATIONS

Destinations presents new directions for both tourism and cultural landscape studies in geography, crossing the traditional boundaries between the research of geographers and scholars of the tourism industry.

Drawing on selected research from Europe, Southeast Asia, the Pacific and North America, the contributors combine perspectives in human geography and tourism to present cultural landscapes of tourist destinations as socially constructed places, examining the extent and manner by which tourism both establishes and falsifies local reality.

Explaining how geographic perspectives about tourist destinations reveal the experience of place for people who live and work in these communities, and examining the forces involved in state planning and global economic activities, the authors show that tourism is essentially about the creation and reconstruction of geographic landscapes through manipulations of history and culture. The "destination," as configured in the tourist's mind, thereby differs from the "actual."

Destinations addresses many critical themes which recent critiques in tourism studies focusing on the attitudes and behavior of the tourist and on the industry as agents of social change have ignored, including the marginalization of the "host" community, the privatization and commodification of local culture, and how tourism acts as both agent and process in the structure, identity and meaning of local places. The authors reveal how geographic conceptualizations of tourist landscapes can constructively anticipate the range of changes wrought on emergent destinations.

Greg Ringer is Adjunct Assistant Professor of Tourism Planning, Public Policy and Management Program/International Studies Program, University of Oregon, USA.

ROUTLEDGE ADVANCES IN TOURISM
Brian Goodall and Greg Ashworth

1 THE SOCIOLOGY OF TOURISM
Theoretical and Empirical Investigations
Edited by Yiorgos Apostolopoulos, Stella Leivadi and Andrew Yiannakis

2 CREATING ISLAND RESORTS
Brian King

3 ECONOMICS OF TOURISM
M. Thea Sinclair and Mike Stabler

4 DESTINATIONS
Cultural Landscapes of Tourism
Greg Ringer

DESTINATIONS

Cultural landscapes of tourism

Edited by Greg Ringer

London and New York

First published 1998
by Routledge
11 New Fetter Lane, London EC4P 4EE

Reprinted 1999, 2000.

Simultaneously published in the USA and Canada
by Routledge
29 West 35th Street, New York, NY 10001

Routledge is an imprint of the Taylor & Francis Group

© 1998 Selection and editorial matter, Greg Ringer; individual
chapters, the contributors

Typeset in Garamond by Routledge

**Printed and bound in Great Britain by
T.J.I. Digital, Padstow, Cornwall**

British Library Cataloguing in Publication Data
A catalogue record for this book is available from the British Library

Library of Congress Cataloging in Publication Data
Destinations: cultural landscapes of tourism/edited by Greg Ringer.
p.cm. – (Routledge advances in tourism; 3)
Includes bibliographical references and index.
1. Tourist trade. I. Ringer, Gregory D., 1951–. II. Series.
G155.A1D478 1998
338.4´791–dc21   97–44289 CIP

ISBN 0–415–14919–3

Tell me where you live and I will tell you by whom you are made!

Anonymous

from THE THIEF

What is it when your man sits on the floor
in sweat pants, his latest project
set out in front of him like a small world, maps
and photographs, diagrams and plans,
everything
he hopes to build, invent or create,
and you believe in him as you always have,
and he hears you, calling him away
from his work,
the angled lines of his thoughts,
into the shapeless place you are bound
to take him,
into the soft geometry of the flesh, the earth
before its sidewalks and cities,
its glistening spires,
stealing him back from the world he loves
into this other world he cannot build without
you.

Dorianne Laux

CONTENTS

Notes on contributors   ix
Acknowledgments   xiii

Introduction   1

PART I
Writing the tourist landscape   15

1 Tourism and the semiological realization of space   17
   GEORGE HUGHES

2 Cybertourism and the phantasmagoria of place   33
   CHRIS ROJEK

PART II
Destinations   49

3 Landscape resources, tourism and landscape change in   51
   Bali, Indonesia
   GEOFFREY WALL

4 Tourism employment and shifts in the determination of   63
   social status in Bali: the case of the "guide"
   JUDITH CUKIER

CONTENTS

5 Rewriting languages of geography and tourism: 80
   cultural discourses of destinations, gender and
   tourism history in the Canadian Rockies
   SHELAGH J. SQUIRE

6 Tourism and the construction of place in Canada's   101
   eastern Arctic
   SIMON MILNE, JACQUELINE GREKIN AND SUSAN WOODLEY

7 Tartan mythology: the traditional tourist image of Scotland   121
   RICHARD W. BUTLER

8 Making the Pacific: globalization, modernity and myth   140
   C. MICHAEL HALL

9 The social construction of tourist destinations: the process   154
   of transformation of the Saariselkä tourism region in
   Finnish Lapland
   JARKKO SAARINEN

   Index   174

CONTRIBUTORS

Richard W. Butler (R.Butler@surrey.ac.uk) is a Professor in the Department of Management Studies at the University of Surrey. He is the author of numerous articles and publications, and co-author of "In search of common ground: reflections on sustainability, complexity and process in the tourism system," *Journal of Sustainable Tourism* (1995) and *Tourism and Sustainable Development: Monitoring, Planning, Managing* (1993).

Judith Cukier (jcukier@waikato.ac.nz) is coordinator of the Tourism Studies Programme lectures in the Department of Geography at the University of Waikato, New Zealand, and has written on tourism and community inter-actions in Southeast Asia. Her writings include "Tourism employment in Bali: trends and implications," in R. Butler and T. Hinch (eds) *Tourism and Indigenous Peoples* (1996) and "Tourism employment: perspectives from Bali," *Tourism Management* (1993). In addition, she co-authored "The involvement of women in the tourism industry of Bali, Indonesia," *Journal of Development Studies* (1996) and "Tourism employment in Bali: a gender analysis," *Tourism Economics* (1995).

Jacqueline Grekin is a Research Associate with the McGill Tourism Research Group, Montreal, Quebec, Canada. She is co-author of "Toward sustainable development: the case of Pond Inlet, NWT," in R. W. Butler and T. Hinch's *Tourism and Indigenous Peoples* (1996).

C. Michael Hall (cmhall@commerce.otago.ac.nz) is a prolific author and editor whose works include *Tourism in the Pacific Rim: Development, Impacts and Markets* (2nd ed., 1997) and *Tourism and Politics: Policy, Power and Place* (1994). He also co-edited *Heritage Management in Australia and New Zealand: The Human Dimension* (1996) and *Tourism in the Pacific: Issues and Cases* (1996). He is currently at the Centre for Tourism, University of Otago, Dunedin, New Zealand and visiting Professor at the Centre for Tourism, Sheffield Hallam University.

George Hughes (George.Hughes@geo.ed.ac.uk) teaches geography at the University of Edinburgh. Previously, he held an appointment in strategic planning and research with the Scottish Tourist Board before joining the Department of Urban Design and Regional Planning and then the Geography Department, where he studies tourism policy and is currently working on a comprehensive analysis of the cultural geography of tourism. His writings include "Authenticity in tourism," *Annals of Tourism Research* (1995) and "The cultural construction of sustainable tourism," *Tourism Management* (1995).

Simon Milne (smilne@ait.ac.nz), Assistant Professor of Geography at McGill University, Quebec, and Professor of Tourism with the Faculty of Commerce, Auckland Institute of Technology, is the author of the forthcoming "Tourism and sustainable development: exploring the global–local nexus," in *Tourism and Sustainable Development*, edited by C. M. Hall and A. Lew; "Travel distribution technologies and the marketing of Pacific microstates," in *Pacific Tourism* (1996), edited by C. Michael Hall and S. Page; and co-author of "Distribution technologies and destination development: myths and realities," in K. Debbage and D. Iaonnides' *The Economic Geography of Tourism* (forthcoming) and *Tourists to the Baffin Region: 1992 and 1993 Profiles* (1997).

Greg Ringer (gringer@oregon.uoregon.edu *or* drtourism@aol.com) is an Adjunct Assistant Professor of tourism at the University of Oregon, USA, and a Visiting Professor in the Departments of Geography and Forestry at Makerere University, Uganda. With experience in sustainable ecotourism and indigenous communities in Africa, Latin America, East Asia and the Pacific, he is the author of "Beyond the boundaries: social place in a protected space," *GeoJournal* (in press) and "Wilderness images of tourism and community," *Annals of Tourism Research* (1996), as well as an upcoming manuscript entitled *Ghosts in the Wilderness: A History of Tourism and Place in Alaska*, for the University Press of Kansas, and another on nature tourism and the political ecology of East Africa.

Chris Rojek (chris.rojek@ntu.ac.uk) is Professor of Sociology and Culture in the "Theory Culture and Society" Research Centre, Faculty of Humanities, Nottingham Trent University. His writings include *Decentering Leisure* (1995), *Ways of Escape: Modern Transformations in Leisure and Travel* (1993), and *Capitalism and Leisure Theory* (1985).

Jarkko Saarinen (jarkko.saarinen@oulu.fi), a Tourism and Recreation Researcher in the Department of Geography, University of Oulu, and with the Finnish Forest Research Institute's Rovaniemi Research Station, has written several articles on tourism in Finland. Among the more recent publications are "The emergence of the tourism region: two approaches to the development of tourism regions," *Terra* (1995) and "The research of social carrying capacity in conservation and recreation areas," *Folia Forestalia* (1994).

Shelagh J. Squire (ssquire@ccs.carleton.ca) is an Assistant Professor in the Department of Geography at Carleton University, Ottawa, Canada. Her research and publications are in the area of tourism and cultural communication, women and tourism history, literary geography, and interdisciplinary studies. Among her publications are "Travels in interdisciplinarity: exploring integrative cultures, contexts and change," *Issues in Integrative Studies* (1995), "Accounting for cultural meanings: the interface between geography and tourism studies re-examined," *Progress in Human Geography* (1994), "The cultural values of literary tourism," *Annals of Tourism Research* (1994), and "Gender and tourist experiences: assessing women's shared meanings for Beatrix Potter," *Leisure Studies* (1994).

Geoffrey Wall (gwall@watserv1.utwaterloo.ca) is a Professor of Geography at the University of Waterloo, Canada, and a prolific author on tourism development and management. His works include "Tourist attractions: points, lines and areas," *Annals of Tourism Research* (1997) and "Indonesia: the impact of regionalization," in F. M. Go and C. L. Jenkins (eds) *Tourism and Economic Development in Asia and Australasia* (1997). He is also the co-author of "Balinese homestays: an indigenous response to tourism opportunities," in R. Butler and T. Hinch (eds) *Tourism and Indigenous Peoples* (1996), and co-editor of *Tourism and Sustainable Development: Monitoring, Planning, Managing* (1993).

Susan Woodley is a Research Assistant with the McGill Tourism Research Group, Montreal, Quebec, Canada.

ACKNOWLEDGMENTS

The authorship of any book is difficult at best, and even more so when it is an edited volume of contributors from a number of countries and institutions. In addition to the complications of coordinating the work of writers who reside in disparate locations around the globe—and whose frequent practice of the art of travel made for a most challenging editorial experience—there is the rather formidable task of integrating a broad selection of professional perspectives and geographic case-studies into a single, coherent manuscript without diminishing the broader diversity of opinion presented. Indeed, so fractious is the complexity of egos and conflicting priorities in similar situations that such endeavors often become meaningless and even impossible.

Yet the eleven contributors to this volume made what might have been an onerous activity one of the most pleasant and informative collaborations in which I have participated. Despite their tight deadlines and the lack of any compensation other than professional acknowledgment and my own sincere appreciation, each writer accepted the challenge presented without hesitation: to encourage a broader discourse of tourism that acknowledges the "place" of the tourist destination in the cultural landscape. To each of these geographers and scholars, I am most thankful.

In addition to their contributions, I must gratefully acknowledge the tremendous aid provided by others who, quite literally, helped me achieve my own destination (and through it, a point of origin) with the completion of this manuscript. Perhaps most importantly, my editorial assistant Ina Zucker, who volunteered her time and energy to assist in the reviews, and to maintain the balance of perspectives I sought. Equal kudos is due to Nick Kohler and Jill Radek, who supplied ample discourse and camaraderie during my research and writing.

In this regard, they were ably assisted by a caring group of international "discussants," including Aubrey Hord, Denise Jackson, David Keeling, Kerry KenCairn, Carla and Dan Dugas, Coleen Fox, Holly Freifeld, Harry Bondareff, Sarah Shafer, Laura Pratt and Karen Lewotsky in the USA; Daudi Kaliisa and Dann Griffiths (Uganda); Yem Sokhan, Paulin Im, Karen and Alan Robinson (Cambodia); Elena del Castillo Perez, Tina Forster and Ernesto Yallico (Peru);

and Laura Gomez and Carlos Soza Manzanero (Guatemala). Together with my parents, family—Barbara, Geoff, Gary, Gordon and Darcie—and mentor, Al Urquhart (Professor Emeritus, University of Oregon) this close circle of friends lent support and meaningful insight whenever needed, enabling me to persevere in the blending of place, practice and philosophy I desired.

Finally, I wish to acknowledge the initial encouragement given by Carolyn Cartier, and most especially my gratitude to the series editor at Routledge, Valerie Rose. Val's personal rapport, patience and responsiveness assured this manuscript's successful transition from concept to finished publication. Though others were certainly involved in its production, *Destinations* exists because these colleagues cared enough to participate in its creation. I am gratefully indebted to each one.

INTRODUCTION

Tourism's exponential rate of growth in the 1990s—one that has proponents now proclaiming it the world's largest service industry—has been stimulated by the booming interest in travel as a viable economic alternative for communities and the trend toward more serious environmental conservation in countries around the world (WTO 1997). It equally reflects the increasing desire among people to engage in meaningful, interactive experiences with local people in other communities and cultures. Certainly, there can be no denying tourism's potency—cultural, economic and environmental—thus presenting possibilities that could not otherwise exist. Yet careful consideration must be given to its development as well, for the social consequences of improper tourism may result in denigration of the very landscape which attracts visitors and, more importantly, sustains the local populace.

While recent critiques correctly focus on the tourist and the industry as agents of social change, several critical themes remain marginalized in the literature of tourism geography and leisure studies, including the commercialization and privatization of local places, and the commodification of the "host community" through the production of tourist landscapes and services (Rojek 1993; Urry 1990, 1994). Certainly, tourism is sufficiently credited with the preservation of cultural heritage and the revival of ethnic identity in select studies, but its effect on the social construction of the tourist destination—though few authors deny the cultural proclivity of tourism—is more likely to be considered tangential or characteristic of the "billiard ball model" Wood describes (1980: 565; cited in Hitchcock et al. 1993: 8; Cosgrove and Daniels 1988; Shaw and Williams 1994; Wild 1994).

The reality, however, is that tourism is a cultural process as much as it is a form of economic development, and the destination of the tourist and the inhabited landscape of local culture are now inseparable to a greater degree (Ingold 1994). By continuing, instead, to treat tourism as exogenous rather than place-centered and constructed, geographers and other social scholars risk ignoring the extent and manner by which tourism both establishes and falsifies local reality.

The development and management of tourism play a significant role in the

1

lives of people who intersect with or live in the destination, and mediates the formation of local identities and cultural patterns of behavior and communication. Through their attitudes and activities, visitors, residents and the tourism industry work out values and goals in destination communities to create inscribed landscapes and places that represent how people interpret and define the experience of living through tourism. Thus to conceive of the cultural destination as a stylized vignette of local history, rooted in time and space, and lacking the dynamic conditions necessary for change, is to render mute the actions, motivations and values of local participants in the ongoing social construction of their place.

The acknowledgment of locally constructed and identified landscapes on the other hand, makes clear how destination communities both adapt to and modify tourism in such a way that it is no longer easily divorced from the cultural mainstream. Rather, the crafting of what Picard calls identifiable, cohesive "touristic cultures" (cited in Hitchcock *et al.* 1993: 8–9, Brightman 1995), gives evidence to the localized morphology and meaning of destinations for both resident and visitor.

It is this rather broadly defined, humanistic and holistic interpretation of tourism and cultural landscape, that this volume offers in crossing the boundaries between the research of geographers and that of tourism scholars. For professionals and educators in tourism and leisure studies, the collected works expand the current literature by emphasizing a localized, socially constructed environment often overlooked; simultaneously, they address the lack of study on tourist destinations in the cultural landscape work of geographers, who until recently concerned themselves primarily with descriptive studies of movement and the broader utilization of space, rather than the recognition of tourist places and "the understanding of the social processes that lie behind these spatial configurations" (Mowl and Towner 1995: 103; see also Shaw and Williams 1994; Squire 1994; Theobald 1994). The result has been the decline in both quality and availability of knowledge and expertise at the community level, even though tourism, as a process of movement as well as "a form of modern consciousness" (Nuryanti 1996: 249) and a social phenomenon—as "a form of cultural policy as much as an economic or industry one" (Craik 1995: 87) is best understood within the context of the human environment where it exists.

With this intent, the contributors to this edited volume seek to conceptualize destination communities as groups of people and their places of lived experience, whose cultural landscapes and local economies increasingly exhibit the influx of new ideas and changing patterns of social interpretation and communication associated with tourism's progress. By so doing, the book suggests alternative geographies of tourism and landscape that transcend the contrived versions of culture and history presented by the tourism industry, to more accurately describe the dynamic places of local people and their visitors, thereby focusing attention on the varied levels of analysis adopted by each group, and perhaps more importantly on the complexity of tourism and the relationship between those levels.

Drawing on a distinct selection of research from Europe, Southeast Asia, North America and the Pacific, attention is focused on tourism as both agent and process in the structure, identity and meaning of local places, embedded as they are within larger regions and economies. Each of these destinations contains great cultural heritage and some of the most biologically diverse sites in the world, yet all must confront the stresses of an increasing population and limited resources, antiquated land tenure systems, health epidemics, conflicting gender roles and ethnic warfare. How they respond, and the role of geography in shaping the discourse and social structure of the resulting landscapes, is at the heart of the story.

To initiate the dialogue, George Hughes and Chris Rojek provide the theoretical core for new directions of thought and vocabulary that span geography and tourism studies in their chapters in Part I. In Hughes' examination of tourism symbols and their function in Scotland, and in Rojek's analysis of the complexities of cybertourism, both authors deconstruct and reassemble cultural and economic approaches that stir the language and framework of analysis about "destinations" in theoretically innovative directions. As such, they set the tone for current thinking and original interpretation about cultural landscapes of tourism, and establish the basis for a new vernacular.

The collected chapters in Part II demonstrate how geographic perspectives about landscapes as tourist destinations reveal the experience of place for people who live and work in these communities, and the range of forces in state planning and global economic activities. Building on the possibilities of "destination" as a theoretical component of both history and location, the case studies suggest how geographic conceptualizations of tourist landscapes might constructively anticipate the range of changes wrought on emergent destination communities. "Destination" connotes the agenda and perspective of the tourist and the industry; the "destination community" becomes the inhabited place of local experience.

Language

The geographer Yi-Fu Tuan (1991: 104) suggests that a central task of cultural geographers is "to understand the making and maintenance of place—the paired features, localities, regions, and landscapes that make up the earth's surface." Thus cultural geographers have historically studied "differences from place to place in the ways of life of human communities and their creation of man-made or modified features . . . [with the] major focus on landscape development and the diffusion over space and time of specific cultural features" (Johnston 1983: 79).

Yet there has been a growing sense among scholars that the quality of the recreational experience is more likely to depend on the contextual meanings attached by individuals to the destination setting, "not just in terms of its physical structure, but also its social environment" (Mowl and Towner 1995: 103). Derived from the ability to observe and participate directly in the physical environment, the images that we associate with the immediate surroundings, as resident and visitor, constitute a very powerful social force that exerts a strong influence on the ways in which we behave in—and the values that we hold of—the destination setting: "people . . . have unique views and feelings. Based on these views and feelings, people develop values, and based on these values they make choices. These choices affect the earth and change the landscape" (Kaplan and Kaplan 1989: 93).

Social perception and construction

Certainly, the perception among local people that they live in a distinctive landscape is essential to their creation of community and a well-defined sense of place, a bond most apparent in the sharing of selected mental and physical characteristics of the particular locality in which they reside over time (Caneday and Zeiger 1991; Hirsch and O'Hanlon 1995). This social milieu supplies the contextual framework for collective acts of spatial organization and behavioral experience, whereby residents establish a place of their own making and acquire a localized "way of life" and a "way of seeing" (Yeoh and Kong 1996: 53). Through this "process of becoming" (Pred 1984) "whose form embodies both our history and our present" (Spirn 1990: 32) residents actively participate in the historically contingent reconstruction of place. The result is a diversity of destinations where "traditional practice is reconstituted within new contexts" (Oakes 1993: 47), the assembly of which is conditioned by relations of individual understanding and social behavior in geographically distinct locations (Johnston 1991).

At the same time, response to the desire for cultural authenticity and reality by some visitors, and recreational diversion for others, makes tourism a dominant form of cultural policy in many areas (Craik 1995). Already, the range of possible effects of increasing tourism has been well documented in terms of changes to the socio-cultural characteristics of the destination community, including the introduction of external elements and the disruption of traditional practices and lifestyles (MacCannell 1973; Dann 1993; Evans-Pritchard 1993). Nevertheless, what receives much of the attention in the professional discourse is the size of the "host population" and the physical space it occupies, rather than the impacts of tourism on the inhabited place as expressed in the attitudes, impressions, and subjective connectedness of local people to each other and to their self-defined community (Cosgrove and Jackson 1972; Relph 1976; Norris 1993).

Landscape and place

As geographic concepts, landscape and place become more than an inert stage upon which the destination community exists (Mitchell and Murphy 1991). Instead, to acknowledge the distinctive cultural landscapes of the tourist destination and its residents' sense of "place" is to recognize a bond formed by conditions of human experience at a variety of spatial scales, both physical and emotive (Spirn 1990; Hirsch 1995; Hirsch and O'Hanlon 1995). This geographical "fact," as Ley (1977: 12) asserts, "is as thoroughly a social product as the landscape to which it is attached . . . [Through the inclusion of] natural elements and human constructions, both material and ideal . . . place [serves as] both a center of meaning and the external context of our actions" (Entrikin 1991: 6).

Clearly then, rather than an unidirectional influence as the "billiard ball" model suggests, reciprocity is at the core of our conceptualization of destination as "a negotiated reality, a social construction by a purposeful set of actors" (Ley 1981: 219). Certainly, as Jaarko Saarinen notes in his study of ethnic tourism in Finnish Lapland in Part II, the delineation by indigenous populations—and tourism researchers—between "insider" and "outsider" serves to effectively establish the physical and perceptual boundaries of the local place. Shaped by both cultural and personal refractions, the shared realities are further reflected in the social structure and organization of the destination community. But the relationship is mutual, for "places in turn develop and reinforce the identity of the social group that claims them" (*ibid.*), allowing local people to situate themselves and their actions in both space and time (Lowenthal 1975).

In this manner, the cultural landscape of the tourist destination delivers a powerful historical and geographical image of the immediate area. Indeed, it is for this reason, in part, that Relph (1976) urges us, in his discussion of placelessness and the "Disneyfication" of tourism landscapes, to become more concerned with the identification of, and respect for, such places, contending that "improved knowledge of the nature of place can contribute to the maintenance and manipulation of existing places and the creation of new places" (1976: 44). To do so requires that the destination be understood, not only in terms of physical and functional attractiveness, but more importantly as a phenomenon of personal experience.

Experience and process

Recognition of the socially constructed destination illuminates and helps to clarify the individualized landscapes of tradition and subjective attachment perceived by people in their own place. Through such representations, a wealth of social and psychological information, both informed and sensed, is revealed regarding the destination as attraction and habitat, and the emotional degree to which that space is individually humanized through direct experience and intimacy.

> In the course of generating new meanings and decoding existing ones, people construct spaces, places, landscapes, regions and environments. In short, they construct geographies. . . . They arrange space in distinctive ways; they fashion certain types of landscape, townscape and streetscape. Human geographies are under continuous invention and transformation by actions whose underlying fields of knowledge are themselves recreated through geographic arrangements. Peoples' cultures and their geographies intersect and reciprocally inform each other . . . in process, in time.
>
> (Anderson and Gale 1992: 4–5)

As such, the social processes of cultural construction are inherently geographic in nature, and the varied intersections of scale and intensity have obvious consequences for sustainable tourism planning and management, as well as the sustainability of the destination itself as attraction and abode (Hunter 1995). Geographers and tourism scholars must therefore become sensitive to the multiple realities of social groups residing in diverse places and developing under different physiographic regimes and historic conditions, who experience, interpret and articulate the values and meanings of their destination in a variety of different ways.

Destinations

Ascertaining the value of the cultural landscape, as expressed and understood through a multidisciplinary perspective of diverse tourism destinations, provides a useful point of entry in examining the ways by which local sensibilities are incorporated into the process of understanding tourism development. While the concept of "landscapes" is somewhat ambiguous, it is an important one for its attention to the manner in which the visible structure of a place expresses the emotional attachments held by both its residents and visitors, as well as the means by which it is imagined, produced, contested and enforced (Eckbo 1975; Lowenthal 1975; Jackson 1984; Gupta and Ferguson 1992).

To the physical geographer, the interplay between human action and geologic process is embedded in the geomorphological artifacts of the landscape and dimensions of space and time. The human geographer, on the other hand, interested in cultural details, must go beyond physical landforms, soils, vegetation and climate, and seek evidence of social interaction in the historic morphology of a place, and the quality and creativity of its cultures. Intent on the study of the organization of space and the random patterns of human activities, the latter is thus apt to conceive the landscape

6

as a living map, a composition of lines and spaces not unlike the composition which the architect or planner produces, though on a much vaster scale . . . the product of innumerable private decisions and inspired by a variety of motives—economic, esthetic, technological, political.

(Jackson 1977: 66)

It is this socially constructed landscape that is of most interest to the authors in this manuscript—the cultural landscape that develops locally as a direct result of the influence of human institutions and values on the physical environment over time (Cosgrove 1984; Duncan and Savage 1989; Harvey 1989; Blomley 1994). At once ambiguous, attractive and important, the visible landscape reflects the human values and ideologies of the resident and viewer, serving as a palimpsest of place attachments between individuals and social groups to specific locations and places (Foucault 1973; Relph 1976; Jackson 1984). As we become more observant of this inhabited destination, spatial and temporal scales enlarge, leading to a process of self-discovery and identification that tourism "takes place in a *place* grounded in information and experience" (Snyder 1990: 39).

In contrast to the contrived and idealized landscape of tourism, the formation of the destination by local people reflects, in the contoured spatial patterns of their settlements and social practices, the peculiarities of the place and those who settle there: "These act as constraints upon present-day processes, both physical and social, and it is important to understand their role. Inherited forms . . . have sometimes been produced in contexts different from those in which modern processes operate" (Wagstaff 1987: 5). By adopting this perspective, the cultural landscapes of the tourist destinations are identified and appreciated as historically dynamic places in perpetual evolution, shaped by social values, attitudes and ideologies as they contract and expand, deteriorate and improve over time and space (Baker and Biger 1992):

to view the landscape . . . is to acknowledge its cumulative character; to acknowledge that nature, symbolism, and design are not static elements of the human record but change with historical experience; and to acknowledge too that the geographically distinct quality of places is a product of the selective addition and survival over time of each new set of forms peculiar to that region or locality.

(Conzen 1990: 4)

As part of our consciousness and as part of our lived realities, tourism is essentially about the creation and reconstruction of geographic landscapes as distinctive tourist destinations through manipulations of history and culture. A transformation of mythology into place, its meaning shaped by multiple contexts of production and consumption, the landscape of tourism is articulated and made visible through the expression and acquisition of experiences, its symbolic nature the product of a linear, social perspective that imparts "a particular way of seeing the world" through travel (Cosgrove 1984: 34; Hirsch 1995: 20). As a result, the means by which tourism is involved in the cultural construction of space through place warrants further attention from geographers and tourism professionals.

Conclusion

In contrast to the traditional literature, the approach advanced here highlights the active role of tourism in creating local geographies through the medium of place. Tourism, it is argued, differentiates space and time in response to the growing globalization and cultural homogenization of the "travel marketplace." This is achieved through means both deliberate and unintentional. Of particular relevance is the gendering of space "through the physical structure and layout of the built environment" (Mowl and Towner 1995: 104; see also Bodenhorn 1993), and tourism's use as an important mechanism to define what is natural or culturally authentic and sustainable (Urry 1990: 98; Lash and Urry 1994).

Clearly, as Geoffrey Wall notes in his chapter on tourism and landscape change in Indonesia, the pace and scale of the changes associated with tourism vary considerably from place to place and from time to time, giving rise to different meanings for different people. Shelagh Squire reiterates, in her discussion of the language of geography and tourism, that destinations—as tourist spaces—are shaped by differences of gender, race, class and ethnicity, as well as the dialectic of leisure and tourism, in addition to their physiographical distinctiveness. Yet only recently has attention been paid to the heterogeneity and social differentiation of visitors and local residents and, as pertinent, the interaction between practical notions of sustainable tourism and those of globalization and economic development, a point elaborated in Judith Cukier's examination of social status and the hierarchy of tourism employment and guiding in Bali.

Furthermore, the "environmental views which we take for granted, or term 'natural,' in our economic exploitation of the earth are not at all natural" (Hughes 1995: 52) but are highly romanticized world views that we hold of travel and the destination as attraction. As such, they reveal vital clues to the dynamic relationships that each of us develops with it, a situation affirmed strongly in the discussion of tourism and the construction of place in Canada's eastern Arctic region reported by Simon Milne, Jacqueline Grekin and Susan Woodley.

In the roles of tourist and "host," we reveal our own cultural and personal world views and biases when introduced to the "new," exposing the socially constructed character of our beliefs and values. The outcome of this process is an industry that satisfies the commercial imperatives of an international business, yet rarely addresses local development needs. While economic precepts assume that the natural and cultural environments of the destination will be commodified as a recreational resource, the significance of the means by which we have commercially constructed nature and ethnicity becomes more explicit when we turn to tourism as a particular economic use of the environment, for

> There is nothing "natural" about the various landscapes valued by tourists since they are the result of the history of human intervention which has manipulated nature in ways conducive to economic growth. . . . Even the experiential character of tourist interaction with the landscape is reduced to defining land as a "tourism resource." What has been washed out of this construction, by the privatizing and domestication of affect, is the domain of culture.
>
> (Hughes 1995: 53)

Yet culture, moderated by local knowledge and institutions, mediates the demarcation of territory and the identity of place (Johnston 1983: 151; see also Gupta and Ferguson 1992). Thus it is an essential interpreter of the constructed landscape of the destination. A theoretical component of both time and space, destination provides the critical context within which visitors and local residents interact, as well as the site where tourism actually occurs. As such, it is both a psychological state of arrival and a process of spatial movement, as well as a culturally defined geography of places, some clearly visible, others not, around which individuals construct and reconstruct ways of life that express "who they are and what is expected of them" (Johnston 1991: 255–6).

It is this broader conceptualization of the geographies of tourism—as place, practice, product and philosophy—that our book applies in its examination of the integrative frameworks through which the destination is shaped through the processes of globalization, modernity and mythology, identified by Michael Hall in his investigation of tourism in the Pacific and in Richard Butler's analysis of Scotland. Applying a critical perspective, we go beyond the rather narrow perspectives of leisure as social control and commodification alone, to recognize a broader range of experience and meaning for those who participate in tourism.

Already, as George Hughes notes in references to the semiology of tourism, the sophisticated institutional structure which made mass tourism socially acceptable now geographically differentiates the globe through the interpretations of travel agents, brochures and books, forcing destinations to establish their distinctiveness as cultural and natural attractions, rather than authentic places of cultural identity. Even more, the possibilities of cybertourism articulated by Chris Rojek, in his qualitative analysis of electronic travel, make clear that conceptions of landscape have now evolved beyond the distinctive, visual spaces of ethnicity, nation and society to include virtual reproductions of social reality, graphically synthesized through radical compressions of both time and place.

Tourism, perhaps more than any other business, is based on the production, re-production and re-enforcement of images (Oakes 1993). These images serve to project the attractiveness and uniqueness of the "other" into the lives of consumers and, if successful, assist in the construction of a network of attractions referred to as a destination. While the complexities of this interaction remain inadequately addressed by tourism geographers and researchers, there is even less study of tourism from an holistic point of view, in which all the elements are conceptually linked, other than from a historical perspective.

Nor is sufficient attention given to the process of planning and implementing sensitive, sustainable community-based tourism programs, or the long-term consequences of evolving cultural tastes on tourism and social form. Thus we are unlikely to achieve any coherent understanding of tourism unless we undertake a broader comprehension of the destination, and what it means to be "native to our places in a coherent community that is in turn embedded in the ecological realities of its surrounding landscape" (Jackson 1996: 3).

Consequently, *Destinations* explores the nature and extent of this symbiotic interaction in response to the challenge facing geographers and tourism scholars to make sense of the often turbulent and highly dynamic social systems within which tourism exists today. The growth of international tourism has made it the dominant service sector industry in the global economy as we approach the millennium, and tourism is now a common prescription for funding sustainable community development worldwide. Many local populations, however, do not yet know how to balance the sometimes opposing goals of economic growth and preservation of tradition and natural environment. Even more, the implications of chaos theory and the "finite nature of perception, observation, and calculation in an infinite world" (Cartwright 1991: 44) makes evident the logical impossibility of assuming predictability in a world where human behavior may be more complex than we presume. As a result, even the most sensitive development and marketing of indigenous landscapes as tourist attractions may transfigure the inhabitants' social history and their dynamics of place in ways still little fathomed.

Though the places and events portrayed in this manuscript refer to specific landscapes and sets of circumstances, the processes described are endemic to the travel industry and affect local communities everywhere. Consequently, the concerns expressed by local residents about social identity, and about the means by which we understand their lives, must be made central to the discourse of tourism if we are to derive any meaningful interpretation of—and support for—the (re)construction of tourism places. Perhaps only in this manner can we make informed judgements about people and the cultural landscapes of the tourist destination.

References

Anderson, K. and Gale, F. (eds) (1992) *Inventing Places: Studies in Cultural Geography*, Melbourne: Longman Cheshire.

Baker, A. and Biger, G. (1992) *Ideology and Landscapes in Historical Perspective: Essays on the Meanings of Some Places in the Past*, Cambridge: Cambridge University Press.

Bender, B. (ed.) (1993) *Landscape: Politics and Perspectives*, Oxford: Berg.

Blomley, N. (1994) *Law, Space, and the Geographies of Power*, New York and London: The Guilford Press.

Bodenhorn, B. (1993) "Gendered spaces, public places: public and private revisited on the North Slope of Alaska," in B. Bender (ed.) *Landscape: Politics and Perspectives*, Oxford: Berg, 169–203.

Brightman, R. (1995) "Forget culture: replacement, transcendence, relexification," *Cultural Anthropology*, 10, 4: 509–46.

Caneday, L. and Zeiger, J. (1991) "The social, economic, and environmental costs of tourism to a gaming community as perceived by its residents," *Journal of Travel Research*, 30: 45–9.

Cartwright, T. J. (1991) "Planning and chaos theory," *APA Journal*, 57, winter: 44–56.

Conzen, M. (ed.) (1990) *The Making of the American Landscape*, Boston MA: Unwin Hyman.

Cosgrove, D. (1984) *Social Formation and Symbolic Landscape*, London: Croom Helm.

Cosgrove, D. and Daniels, S. (eds) (1988) *The Iconography of Landscape: Essays on the Symbolic Representation, Design and Use of Past Environments*, Cambridge: Cambridge University Press.

Cosgrove, I. and Jackson, R. (1972) *The Geography of Recreation and Leisure*, London: Hutchinson.

Craik, J. (1995) "Are there cultural limits to tourism?" *Journal of Sustainable Tourism*, 3, 2: 87–98.

Dann, G. (1993) "Socio-cultural issues in St Lucia's tourism," paper presented to the International Conference on Sustainable Tourism in Islands and Small States, Foundation for International Studies, Malta, 18–20 November.

Duncan, S. and Savage, M. (1989) "Space, scale and locality," *Antipode*, 21, 3: 179–206.

Eckbo, G. (1975) "Qualitative values in the landscape," in E. Zube, R. Brush, and J. G. Fabos (eds) *Landscape Assessment: Values, Perceptions and Resources*, Stroudsburg PA: Dowden, Hutchinson and Ross, 31–7.

Entrikin, J. N. (1991) *The Betweenness of Place: Toward a Geography of Modernity*, Baltimore MD: Johns Hopkins University Press.

Evans-Pritchard, D. (1993) "Ancient art in modern context," *Annals of Tourism Research*, 20, 1: 9–31.

Foucault, M. (1973) *Order of Things: An Archaeology of the Human Sciences*, New York: Random House.

Gupta, A. and Ferguson, J. (1992) "Beyond 'culture': space, identity, and the politics of difference," *Cultural Anthropology*, 7, 1: 6–23.

Harvey, D. (1989) *The Condition of Postmodernity*, Oxford: Blackwell.

Hirsch, E. (1995) "Landscape: between place and space," in E. Hirsch and M. O'Hanlon (eds) *The Anthropology of Landscape: Perspectives on Place and Space*, Oxford: Clarendon Press, 1–30.

Hirsch, E. and O'Hanlon, M. (1995) *The Anthropology of Landscape: Perspectives on Place and Space*, Oxford: Clarendon Press.

Hitchcock, M., King, V. T. and Parnwell, M. J. G. (eds) (1993) *Tourism in South-East Asia*, London and New York: Routledge.

Hughes, G. (1995) "The cultural construction of sustainable tourism," *Tourism Management*, 16, 1: 49–59.

Hunter, C. J. (1995) "On the need to re-conceptualise sustainable tourism development," *Journal of Sustainable Tourism*, 3, 3: 155–65.

Ingold, T. (ed.) (1994) *Companion Encyclopedia of Anthropology: Humanity, Culture and Social Life*, London: Routledge.

Jackson, J. B. (1977) "A new kind of space," in E. Zube and M. Zube (eds) *Changing Rural Landscapes*, Amherst MA: University of Massachusetts Press, 66–73.

——(1984) *Discovering the Vernacular Landscape*, New Haven CT and London: Yale University Press.

Jackson, W. (1996) *Becoming Native to This Place*, Washington DC: Counterpoint.

Johnston, R. J. (1983) *Geography and Geographers*, 2nd ed., London: Edward Arnold.

——(1991) *A Question of Place: Exploring the Practice of Human Geography*, Oxford: Blackwell.

Kaplan, R. and Kaplan, S. (1989) *The Experience of Nature: A Psychological Perspective*, Cambridge: Cambridge University Press.

Lash, S. and Urry, J. (1994) *Economies of Signs and Space*, London and Thousand Oaks CA: Sage.

Ley, D. (1977) "The personality of a geographical fact," *The Professional Geographer*, 29: 8–13.

——(1981) "Behavioral geography and the philosophies of meaning," in K. R. Cox and R. G. Golledge (eds) *Behavioral Problems in Geography Revisited*, New York: Methuen, 209–30.

Lowenthal, D. (1975) "Past time, present place: landscape and memory," *The Geographical Review*, 65: 1–36.

MacCannell, D. (1973) "Staged authenticity: arrangements of social space in tourist settings," *American Journal of Sociology*, 79, 3: 589–90.

Mitchell, L. and Murphy, P. (1991) "Geography and tourism," *Annals of Tourism Research*, 18: 57–70.

Mowl, G. and Towner, J. (1995) "Women, gender, leisure and place: towards a more 'humanistic' geography of womens' leisure," *Leisure Studies*, 14: 102–16.

Norris, K. (1993) *Dakota: A Spiritual Geography*, Boston MA and New York: Houghton Mifflin.

Nuryanti, W. (1996) "Heritage and postmodern tourism," *Annals of Tourism Research*, 23, 2: 249–60.

Oakes, T. S. (1993) "The cultural space of modernity: ethnic tourism and place identity in China," *Environment and Planning D: Society and Space*, 11, 1: 47–66.

Pocock, D. C. D. (ed.) (1981) *Humanistic Geography and Literature: Essays on the Experience of Place*, London: Croom Helm.

Pred, A. (1984) "Places as historically contingent process: structuration and the time-geography of becoming places," *Annals of the Association of American Geographers*, 74: 279–97.

Relph, E. (1976) *Place and Placelessness*, London: Pion.

——(1981) *Rational Landscapes and Humanistic Geography*, London: Croom Helm.

Rojek, C. (1993) *Ways of Escape: Modern Transformations in Leisure and Travel*, London: Macmillan.

Shaw, G. and Williams, A. (1994) *Critical Issues in Tourism: A Geographical Perspective*, Oxford: Blackwell.

Smoliner, C. (n.d.) *Research Initiative: Cultural Landscapes*, Vienna: Bundesministerium fur Wissenschaft, Forschung und Kunst, Refererat II/A3a.

Snyder, G. (1990) *The Practice of the Wild*, San Francisco CA: North Point Press.

Spirn, A. (1990) "From Uluru to Cooper's Place: patterns in the cultural landscape," *Orion Nature Quarterly*, 9, spring: 32–9.

Squire, S. (1994) "Accounting for cultural meanings: the interface between geography and tourism studies re-examined," *Progress in Human Geography*, 18: 1–16.

Theobald, W. (ed.) (1994) *Global Tourism: The Next Decade*, Oxford: Butterworth-Heinemann.

Tuan, Y. (1977) *Space and Place: The Perspective of Experience*, Minneapolis: University of Minnesota Press.

——(1991) "A view of geography," *The Geographical Review*, 81, January: 99–107.

Urry, J. (1990) *The Tourist Gaze: Leisure and Travel in Contemporary Societies*, London: Sage.

——(1994) "Cultural change and contemporary tourism," *Leisure Studies*, 13: 233–8.

Wagstaff, J. M. (1987) *Landscape and Culture: Geographical and Archaeological Perspectives*, New York: Blackwell.

Watson, W. (1989) "People, prejudice, and place," in F. W. Boal and D. N. Livingstone (eds) *Behavioural Environment: Essays in Reflection, Application, and Re-evaluation*, London: Routledge, 93–110.

Watts, M. (1988) "Struggles over land, struggles over meaning: some thoughts on naming, peasant resistance and the politics of place," in R. Golledge, H. Couclelis and P. Gould (eds) *A Ground for Common Search*, Goleta CA: The Santa Barbara Geographical Press.

Wild, C. (1994) "Issues in ecotourism," in C. P. Cooper and A. Lockwood (eds) *Progress in Tourism, Recreation and Hospitality Management*, vol. 6, Chichester: Wiley, 12–21.

Wood, R. (1993) "Tourism, culture and the sociology of development," in M. Hitchcock, V. T. King and M. J. G. Parnwell (eds) *Tourism in South-East Asia*, London and New York: Routledge, 48–70.

World Tourism Organization (1997) *Tourism Highlights 1996*, Madrid: WTO.

Yeoh, B. and Kong, L. (1996) "The notion of place in the construction of history, nostalgia and heritage in Singapore," *Singapore Journal of Tropical Geography*, 17, 1: 52–65.

Part I

WRITING THE TOURIST LANDSCAPE

TOURISM AND THE SEMIOLOGICAL REALIZATION OF SPACE

George Hughes

In 1738, John Wesley, with his brother Charles, founded the movement that became the Methodist Church. Two hundred and fifty years later, in 1988, the British Tourist Authority (BTA), using the 250th anniversary, considered the possibility of establishing a Wesley Trail intended to appeal to an estimated 54 million Methodist worshipers worldwide. That such a conjunction of tourism and evangelism was considered sufficiently uncontroversial to be pursued by a national tourist authority, and to be used subsequently in a vocational tourism text (Lumsdon 1992: 132) is symptomatic of the scale of cultural change that has occurred in the intervening centuries and of the current role tourism is playing in facilitating and reflecting the change.

The objectives established for this project did not address the religious sensibilities of Methodists or conceive of the trail as a mark of respect for the Wesleys or Methodism. Rather, the project was addressed in exclusively commercial terms. More specifically, the idea satisfied a number of criteria in terms of BTA objectives:

1   The Methodist Church provided a clearly-defined group which could be targeted;
2   As Wesley had so many strong links with different parts of the country, such a promotion would spread the benefits to areas which would otherwise not readily attract overseas business;
3   The idea had potential for sound public relations work which would have a wider impact than the specific promotion;
4   The campaign could also be extended to include the centenary of John Wesley's death in 1991;
5   The campaign had a potential for joint sponsorship (Lumsdon 1992: 133).

Yet this potential interaction of tourist development with a belief system, so important in the structure of many Western societies, arose capriciously from the touristic opportunity afforded by an anniversary. The opportunistic, accidental and aleatoric ways in which tourism alights on such themes, and the potential this has for "realizing" particular geographies demands attention if only for its seemingly limitless embrace of topics and places in the world. This spatializing potential raises particular issues and offers an opportunity to consider some of the contrivances at work in the late twentieth century which are subtly but perceptibly changing the cultural and physical geography of the globe.

As tourism impinges on the remoter margins of the world, so it has been increasingly implicated in concerns about the cultural and physical wellbeing of the localities that it incorporates. A brief acquaintance with the academic literature on tourism quickly demonstrates a critical turn in which tourism has been routinely censured for its economic insecurity and its cultural exploitativeness (Crick 1989). The question at issue therefore seems no longer to be one of whether tourism has become a potent agent of change, but of how to address the character of this change. In this chapter I wish to develop a geographical perspective in which tourism will be considered as a spatially differentiating activity which has the potential to realize different "geographies" in a semiological way.

Tourism is, after all, essentially about making available a diverse range of geographical locations to potential visitors and thereby translating those locations into tourist destinations. In other words, tourism is inherently geographical. Yet this proves to be a very ambiguous process, for while the necessity for destinations to be imaginatively differentiated is recognised, in order to capture particular tourist markets, the process has been simultaneously criticized as a culturally homogenizing one. As places get to look more like each other, the rhetoric of tourism has been stretched in its attempt to contrive geographical distinctiveness. Thus the ways in which tourism is involved in the construction of place and space warrants attention from geographers.

The geography of tourism

In the light of the current ubiquity of pleasure, the relative paucity of geographical studies on tourism and leisure seems curious. Geographically, tourism has been treated as marginal to the serious business of both economic and academic production. Geographers have not entirely neglected tourism and recreation, but rather their studies have been predominantly taken up with monitoring and describing the characteristics of recreational land use patterns and flows (Coppock 1982). These patterns and flows have been discussed as if they were an effect of some more fundamental causes, such as the tendency for frequency of visits to decline with the distance of travel (or distance decay), car ownership (and its influence on the propensity to travel), age and stage in the life-cycle (and the motivation to undertake travel) and the like. As such, leisure, recreation and tourism have tended to be treated by geographers as the epiphenomena of a range of more fundamental spatial determinants.

In contrast, this chapter will assert that leisure and tourism, being central to social life, are primary determinants of space in their own right. This has been achieved, semiologically, as places have become subject to strategies of "theming," "designation," "re-visioning," "re-imaging" and the comprehensive application of marketing techniques that are now the familiar repertoire of place marketing. Taken collectively this has generated what may been called "ludic space" (Lefebvre 1976: 82–4) which is now probably the most extensive land use in Western society.

That tourism works through dreams (Reimer 1990) and myths (Selwyn 1990, 1996) is now widely recognized, but the spatializing character of these dreams and myths has been somewhat less developed (but see Goss 1993; MacCannell 1973, 1976) as is also the role of these in differentiating space into places. Lash and Urry (1994), drawing on a larger thesis about the characteristics of modernity, draw attention to the institutionalization of leisure and tourism that was a prerequisite for the development of travel as a mass activity. They argue that the professionalization of travel engendered public trust in its institutions and made it possible for anonymous captains, pilots and couriers to transport and accommodate groups of tourists.

Travelers were likely to be strangers to each other, and their significant personal confinement a necessity of the journey, yet anxieties about personal safety and security have been largely circumvented. But the sophisticated institutional structure, which made mass tourism socially possible, has also been a leading factor in spatializing the globe, since the institutions of tourism are also active in geographically differentiating the world. The trust that the tourist invests in those that arrange for their safe conduct also extends to the character of the places visited, for much tourism relies on the qualities of a destination as represented to tourists through guidebooks, brochures, travel writing, tourist programs and the like.

In tourist promotions, places are represented in a kind of patois. Destinations are referred to, epigrammatically, as "the sunshine coast," "the city of discovery," "a world of a difference," "cultural capital," etc. This is a rhetorical shorthand intended to convey some kind of spatial identity. It is equivalent to what Shortridge (1984) elsewhere, although not in the context of tourism, considered to be a form of regional labeling. These epithets purport to represent a particular set of traits which may be associated with a group of residents and are not, therefore, to be dismissed as trivia. They insinuate dreams and myths into the public perception of places which may come, in time, to stand, like icons, logos or mottoes, as shorthand statements of their character. How this engages with the affective attachment of residents to their own localities is still a matter of some speculation, however, because it has not been sufficiently taken

into account in geographical treatments of the function of place in the formation of group and individual identity (see Boyle and Hughes 1995). However,
although under-reported, the place-representational role of tourism in differentiating space would appear to be culturally substantial.

But place representation is an *active* form of spatializing. The place depicted
is intentionally constructed rather than being a passive outcome of some more
"serious" economic activity such as mining or manufacturing. The geographical character of a tourist destination is aesthetically managed with an eye to
its tourist market potential. This distinction between active and passive
spatial differentiation is an important one, for it distinguishes the traditional
geography of space from a more cultural approach.

Space may be treated in two fundamentally different ways. While for
quotidian purposes it seems impossible to escape the experience of being in
space, it does not axiomatically follow that social behavior will always deploy
spatializing strategies. Tourism, I wish to show, may be studied in the latter
way. Yet this use of place representation should not be interpreted simply as
the latest example of irrepressible capitalist exploitation of the disenfranchised
(although there are multitudinous examples of this), for the rash of re-imaging
strategies that seems to have infected every town and city in the Western
world is just as likely to come from within those communities as to be
imposed on them from without. As their functions have waned, so towns and
cities have become actively engaged in reconfiguring their identity to get
themselves onto the tourist map (Hughes 1992; Short *et al.* 1993).

Whether externally or internally motivated, however, what is particularly
noteworthy is that the semiological differentiation of space has become a
highly self-conscious, self-reflective process. The content and form in which
places are represented can no longer, if they ever could, be thought of as
evolving naturally. Rather they reflect an unstable outcome of a struggle
between interest groups in which the self-conscious intervention of tourism
entrepreneurs, travel writers, governments and tourists have come to play an
increasing role.

Organizing tourist space

Tourism is an important conduit for organizing meanings in space. Such
meanings may "evolve" as explorers and adventurers are succeeded by less exis-
tential forms of mass tourism (Cohen 1973, 1974, 1979); they may be
formally imposed by corporate mandate, as in the development of resort
complexes; they may be the self-conscious product of local community aspirations for economic development (Boyle and Hughes 1995); or, most probably,
some blend of all three. The effect of tourism, however, is to reconfigure the
existing cultural and physical endowment of places as they become installed in
the more autonomous discourse of tourist consumption.

Tourism, as a mode of spatial organization, confers exchangeability on artifacts by reconfiguring country houses, palaces, safari parks, game reserves, coal mines, theme parks, etc., as tourist attractions (Urry 1990; Crang 1994). In so doing, the particularity of the individual histories of each become subordinated to the thematic demands of a touristic mode of consumption. As tourist "sights," they carry "markers" (MacCannell 1973, 1976) whose descriptions are likely to deny the cultural and political depths behind their conception. Boniface and Fowler (1993) speak, for example, of the risks involved in presenting heritage to an international audience of tourists. Such presentations tend to homogenize particularity and in the process corrupt it.

> we believe the new circumstance of A heritage presentation to a world audience comprised of peoples not just from many countries but also of many different cultural components, is an intrinsically perilous situation . . . [and] . . . those who embark upon the presentation of heritage to the global traveller should regard themselves as addressing an extremely difficult and hazardous task.
>
> (Boniface and Fowler 1993: 150)

Yet even the ubiquitous sun, sea and sand continue to assert their spatial particularities in order to attract tourists to one destination over another. This ambiguity between sameness and difference is a useful point of departure from which to examine some of the qualities of tourist space.

Hyper-reality

The notion of space being not real but hyper-real owes much to the work of postmodern authors. It draws attention to the mediating influences which ensure that the representation of a place is never a simple given but is rather a social construction. What makes the ludic space of tourism hyper-real is perhaps easiest to understand in the context of heritage, where the appeal of nostalgia dictates that what is offered to the tourist is authentically restricted. Heritage attractions (as all tourist attractions) must be safe, clean and pleasing. In fulfilling such conditions, each artifact is displaced from its historical moorings and effectively becomes a different object.

Even if based on authentic historical precedents, heritage objects tend to acquire new meanings, which "overwrite" their original significance. In this process the complexities of social life are washed out and replaced with promotional gestures that fuse with the evocative expectations of the visitor. They acquire, in other words, a different reality. Judgmentally, this reality is likely to be evaluated aesthetically rather than scientifically or morally. As a priority it must "look right" irrespective of historical verisimilitude, cultural validity or moral probity. It must be "pretty as a picture."

In attempting to fulfill such expectations, places engage in intensive

21

programs of aesthetic management. City streets have become pedestrianized and "traditional" ironwork street furniture, resonant of Victorian England, "re"-introduced. Buildings are floodlit, lending the night-time streetscape a theatrical atmosphere, and their frontages historicized and conservation policy extended. Reproduction is the password to a successful tourist destination, for it seems now obligatory, if visitor expectations are to be met, for a destination to engage in aesthetic reconstruction.

The geographical expression of this is no less palpable for its being woven out of the fancies of brochure copywriters or travel journalists, but it is a pastiche of the past. It has no historical authenticity and it evokes a sense of a past that never was. These are spaces that are self-evidently real but, in being faithful to no antecedent circumstances, exhibit a heightened reality framed by the systematic collection, in one place, of aesthetically evocative stimuli.

Liminality

Turner (1974) adapts the concept of liminality to describe the "betwixt and between" moments when people are disposed to feel liberated from the norms of their society. Classically, it described the transitional periods associated with a "rite of passage." Shields (1991) deploys the concept to describe the social status of the beach:

> The liminal status of the eighteenth-century seashore as an ill-defined margin between land and sea fitted well with the medical notion of the "Cure." Its shifting nature between high and low tide, and as a consequence the absence of private property, contribute to the unter-ritorialised status of the beach, unincorporated into the system of controlled, civilised spaces. As a physical threshold, a limen, the beach has been difficult to dominate, providing the basis for its 'outsider' position with regard to areas harnessed for rational produc-tion and the possibility of its being appropriated and territorialised as socially marginal.
>
> (Shields 1991: 84)

The beach offers an example, in microcosm, of the capability of tourism to contest the dominant social norms that reign in space. The social acceptability of states of undress and physical intimacy, common on the holiday beach, appear to have been spatially extended beyond the physical confines of the beach. Indeed, although difficult to categorize because of its "abnormality," this very laxity of dress and behavior offers visual clues as to who is a tourist. Being a tourist seems of itself, therefore, to be sufficient justification for a wide range of social behaviors that would otherwise attract social sanction.

The assertion of such liminal practice is thus another defining characteristic of tourist space.

Libidinal

In modern tourism Western capitalism appears to have developed the ultimate consumer. Baudrillard, in *Consumer Society* (Poster 1989) defines consumption as a kind of hysterical flight driven by an insatiable sense of "lack." His attack is upon the traditional economic anthropology of need fulfilment since, for Baudrillard, there can never be fulfilment, or even a definition of need. Consumption, for Baudrillard, is more like fashion than utility. Goods are acquired and disposed of for their image value rather than their content and the "economy of the sign" is driven by an insatiable desire to keep pursuing difference. Setting to one side Baudrillard's inclination to run to excess in his writings, tourism does seem to resonate with his apocalyptic vision, for his hysterical consumer of signs, or what we might term here the "virtual hyper-consumer," has been given form in the shape of the modern tourist.

In reporting tourism it has become conventional to speak of "trips" to make clear that the resultant visitor patterns are the accumulative effect of many tourist visits. However, seen as a kind of "aggregate consumer," each of which occupies a bed space and a restaurant cover for the complete tourist season, we can alternatively conceive of a pattern composed of virtual hyper-consumers, each of which is framed from the accumulated consumption of multiple trips. Conceived in this way, the emphasis is shifted from the volume of tourist consumption toward the form in which that consumption takes place. This approximates to the Baudrillardian enunciation of the hysterical consumer, since the bulk of tourist trips are indeed undertaken with an expectation of excess. To each individual the brevity of the extravaganza is an appropriate counterpoint to the everydayness of the rest of the year, but taken collectively the sequential procession of tourists produces a concentration of what amounts to virtual, libidinally driven hyper-consumers.

This attempt at configuring a virtual hyper-consumer from the fragmented reality of individual tourist visits may appear to be somewhat gratuitous. However, its spatial significance rests on the repetitiveness of behavior, which despite being enacted through many individuals, bodies itself forth in sufficiently stable performances to be serviced as consumers of distinctly different tourist products, such as type of accommodation and activities. Thus does the multitude of tourists resolve itself into "hotel guests," "caravanners," "backpackers," "sightseers" and the like, which require administration through daily service routines that must be sustained "above and beyond" each tourist as an individual visitor. In this fashion is the behavior of these virtual hyper-consumers inscribed on the environment as entrepreneurs (or "hosts") respond to their expectations as markets (or "guests"). This view of space is a meta-level one framed from sustained hyper-consumption of largely intangible objects under social conditions that are expressly "abnormal." It is little wonder then that tourism has become an archetype of social change for postmodern authors such as Jean Baudrillard and Umberto Eco.

Producing tourist space

But the characteristics of ludic space are, as discussed above, at least partly a product of interventionist strategies that are intended to be spatially differentiating. To illustrate some of the character of this active spatialization, I will examine four genres used in the broadcast of tourism. In each case, it will be evident that the genre, as delivered, has responded to a paradigmatic struggle over the conventions which govern its design. In reporting a variety of opinions regarding good design and its functional efficiency, my objective has been to draw attention to the degree of autonomy that attaches to it. Each, in other words, is heavily immersed in the discourse of its own production and this partly controls the subject matter and how it comes to be presented. Hence the manner in which local "geographies" may be semiotically realized, or produced, becomes subservient to the design conventions which currently control each genre.

The guidebook

Ousby (1990) suggests that the first reported use of a guidebook, in the nineteenth century, was in Byron's poem "Don Juan." However, they were available earlier than this, as evidenced by the established popularity of William Wordsworth's *Guide Through the District of the Lakes*, published first as an appendix to another work as early as 1810 and then as a separate book in 1822 (Wordsworth 1906). Some would trace the ancestry of the guidebook back to the Greek writer Pausanius' detailed account of his travels in the Eastern Mediterranean in the second century A.D. (Deller and Stoelting 1990). Others contend that the first recorded travel books, such as Marco Polo's twelfth-century *Divisament dou Monde* and *The Travels of Sir John Maundeville*, began to circulate in Europe between 1356 and 1366 (Adams 1962).

Whatever their origin, guidebooks succeeded the early use of local residents as "guides" or informants. However, the unreliability and uncertainty of residents' local knowledge encouraged the systematic development of the guidebook, which had matured into something like its current format by the nineteenth century. The guidebook did not therefore come into the world fully formed but has had its particular historical gestation, reflecting the actions of publishers, changes in printing and publishing technology, and the development of market opportunities opened by increasing affluence and improved communication.

John Murray is credited with establishing the modern guidebook for the

English-speaking traveler, with organized itineraries, summary descriptions of sites, evaluations of facilities, indexes, in a pocket size and with regular revisions. Baedeker, who was in competition with Murray, introduced the now widely copied "star system" for rating the importance of tourist sites. It is alleged that Herman Goering instructed the German air force to bomb every historical site and landmark in England that had an asterisk in the Baedeker guide. Hence the infamous "Baedeker Raids" of World War II (Otnes 1978).

The main service that the guidebook renders visitors is a reduction in the complexity of choices about things to do, things to see and places to eat and sleep. It is intentionally evaluative and therefore constitutes a particular kind of geographical text. But guidebooks are distinguishable from academic geographical texts. Their language tends to be evocative and replete with references to the picturesque, spectacular, bustling, glittering, foreboding, majestic, dramatic, colorful, rustic, imposing or charming. In Scotland, kilts "twirl" and Loch Lomond's banks are always "bonnie." Yet in an attempt to avoid accusations of being kitsch or romanticized, some, usually the more recent publications, have attempted a kind of new realism which comes close to academic production.

Like the Irish, the Scots have realized that there is money to be made from conforming to a stereotypical image, however bogus it may be.

> Although few in Scotland ever eat haggis (once described as looking like a castrated bagpipe), it is offered to tourists as the national dish. Heads of ancient Scottish clans, living in houses large enough to generate cash flow problems, have opened their homes to tour groups of affluent Americans. Others have opened "clan shops" retailing an astonishing variety of tartan artifacts, including "Hairy Haggisburger" soft toys. Still others have taken to appearing in Japanese TV commercials extolling their "family brand" of whisky. The cult of the kilt—based, someone mused, on the self-deception that male knees are an erogenous zone—is a huge commercial success.
>
> (Bell 1995: 23)

This self-conscious reflective approach is more difficult to distinguish from the formal academic text but can be done so by considering the audience the text addresses. The targeted reader is not an academic but a tourist. The evocative stance engages with a discourse of value for money and quality of experience in ways unfamiliar in the objective discourse of an academic text.

> But the image obscures the real Scotland. It's worth lingering long enough to draw back the tartan curtain and get to know one of Europe's most complex peoples.
>
> (Bell 1995: 23)

Ultimately, therefore, the guidebook is an invitation to a *performance* or engagement with a place that, however superficial, is different than the abstraction sought from the objectivity of the formal academic text. This performance is arranged through narrative structures. First, the reader is stimulated, by panoramic photographs and/or cultural curiosities. The "plot" is then developed through the use of historical summaries and cross-sections of regional areas. And finally the denouement resolves events, usually concluding the historical review by arriving in the present—in the case of the above, at "Scotland today." The link between the text and the place is that this all "maps" onto reality by the contrivance of spatially arranged itineraries of sites. The performance is "closed" with the tourist's completion of the itinerary, although diversions from the "well trodden" are also frequently offered.

The advertisement

Due to the physical limitations imposed on its text, the advertisement works somewhat differently than the guidebook. Advertisements come in different media and appear in different contexts. Because of their significance to the buoyancy of the commodity market they have attracted considerable market research, and only limited reference is possible here, which will be confined primarily to the printed advertisements of magazines and newspapers. Starch Inra Hooper Inc. publish three types of scores (starch tests) on the readership of advertisements. They classify "noted," "seen-associated" and "read most" (Rossiter 1988). These classifications relate to the percentage of magazine readers who: remember seeing a particular advert in a specific issue (noted); saw or read part of the advert which includes product or company name (seen-associated); and read 50 percent or more of the advertising copy (read most). The scores are widely assumed to relate to the success of the advertisement although this is not undisputed (Holbrook and Lehmann 1980; Rossiter 1981a; Finn 1988).

Readership appeal may also be correlated with information processing models. Thus Finn (1988) proposes that the information processing stages of "exposure," "attention," "comprehension" and "elaboration" may be individually linked to characteristics of the design of an advertisement, such as its size, location and print vehicle (exposure), the layout and dominant illustration (attention), secondary elements such as a logo or headline (comprehension) and the detail of the advertising copy used (elaboration).

At a yet more detailed level, the layout of the advertisement is presumed to have a powerful influence on attention. One study considered the effect of variations in layout, such as block and non-block, illustration and no illustration, different language styles and sentence structure (Motes *et al.* 1992). These factors were analyzed for their influence on the advertisement's appeal, believability, clarity, attractiveness and informativeness, as well as the overall reaction and likelihood of reading the advertisement and using the service advertised.

Other research emphasizes the importance of a dominant focal center, such as a photograph (Starch 1966) and the size and color of the advertisement (Valiente 1973). From information such as this, Ogilvy (1983) has attempted to formulate layout rules for advertisements, suggesting for example that pictures should always be at the top of the page with the headline underneath. Rossiter (1981b) suggests that visuals containing objects, (people, scenery and the like, rather than abstract images), interactive visuals (which juxtapose a user with the product) and increased pictorial content in an advertisement lead to increased readership attention.

Given this attention to the technology and content of advertisements, it is hardly surprising, then, that a systematic construction of the geography of places might be identified in tourist advertisements. Britton (1979), examining the image of the Third World in tourism marketing, attended to "distortions" in the depiction of places as paradisaic, unspoiled and sensuous. Goss (1993) undertakes to reveal similar constructions in his examination of the tourist advertising of the Hawaiian Islands.

He identifies five components, which he calls topoi, by which Hawaii is spatialized in advertisements. It is presented as an earthly paradise, drawing on allusions to the Garden of Eden (embellished with references to the green, lush, verdant, rich and fertile environment) with an eloquent dose of sensuality. Hawaii is marginalized, through mystification of its location "in the middle of the deep blue sea," and culturally as a "chop suey of cultures" which can offer comprehensive experience of international cuisine. It is liminal in its presentation as a threshold between past and present, nature and civilization, east and west and sea and land, and feminized by allusions to penetration and the sensuality of native culture. Finally, Hawaiian advertising offers the promise of "aloha" which has been disembedded from its original cultural context and offered as hospitality, friendliness and magic.

The tourist brochure

The tourist brochure has also attracted the analytical attention of geographers. Dilley (1986) considered the international travel brochure to be

> one of the most important—and often the most important—medium for the formation of images of overseas recreational opportunities . . . the material used . . . is the closest thing to an official tourist image of each country: whatever image the tourist may have, whatever image some third-party company may wish to promote, this is how the countries themselves wish to be seen.
>
> (Dilley 1986: 64)

Dilley applied a content analysis to the international brochures of twenty-one countries and undertook a frequency count of their illustrations, which were classified into landscape themes, culture themes, recreational themes and services. Each illustration was then further examined for the messages conveyed, and the importance of these—as recorded by the frequency of the recurrence of the message—was weighted by the size of the illustration. From this, certain regional images emerged with, for example, the Caribbean featuring water and beach but Old World destinations featuring history, art and architecture.

Wicks and Schuett (1991) tackle what proves to be a more indirect aspect of the influence of the brochure on the construction of places. Their analysis was designed to offer the trade some findings on the effectiveness of brochure dissemination rather than brochure content, yet as they remark, the travel brochure represents an opportunity to provide a great number of strong selling messages (or what is being interpreted here as place-making motifs). Nor need these messages be as ephemeral as the ubiquity of the tourist brochure might imply.

> Assuming the recipient/consumer keeps the promotional piece, it may serve as a long-term reminder or reference. Furthermore, the brochure may be passed on to others, thus multiplying its effectiveness.
>
> (Wicks and Schuett 1991: 302)

In general, Wicks and Schuett found that travelers who had requested brochures felt the material that they had received was "complete" and "well suited" for their travel plans. Selwyn (1990), by contrast, undertakes an anthropological "reading" of some international tourist brochures, using the categories of social attachment and social solidarity, liminality, culture and language more in concert with the work of Goss (1993) and Britton (1979). For Selwyn, brochures operate like myths, the ingredients of which are instantaneity (gratification from sex and food), boundary fragmentation (where high and low culture collapse, places become decontextualized as blue seas and sandy beaches, and the meaning and value of things becomes leveled), individual "sovereignty" (with the market power of choice and "good value" linked to the fantasy power of omnipotence—"king for a day") and schizophrenia (in which the only link between sites and experiences is the deployment of superlatives such as breathtaking, spectacular and awe-inspiring).

Some research into brochure design has also been undertaken by tourist organizations, but most of this remains unpublished. One piece of available research is a study of Thomson Holidays' main brochures (Riley and Spackman 1978) in which the copy length, number of hotels per page and price information are evaluated for their effect upon "desirable" features, although the latter are not detailed in the text. On a more practical level, the English Tourist Board publishes guidelines for print producers (English

Tourist Board 1987) and these make recommendations regarding the use of "sensible" typefaces, concrete and truthful language, cross-heads and the captioning of photographs.

Tourist trails

The tourist trail has become an established part of the portfolio of tourism development and marketing instruments. After a period of considerable innovation in the 1970s, the rising costs of trail management have posed dilemmas for financially hard-pressed authorities, yet there is evidence that themed trails continue to be a popular form of development. Silbergh *et al.* (1994) define a themed trail as

> a route for walking, cycling, riding, driving or other forms of transport that draws on the natural or cultural heritage of an area to provide an educational experience that will enhance visitor enjoyment. It is marked on the ground or on maps, and interpretative literature is normally available to guide the visitor.
>
> (Silbergh *et al.* 1994: 123)

In 1978 there were estimated to be over 800 trails in existence in the UK (Dartington Amenity Research Trust 1978). The widespread use of these trails thus invites analysis as part of an awareness of the culture-forming potential of tourism in general and its spatial manifestation in particular. Of the four genres discussed, it is the tourist trail that has the most direct relationship between the organizing principle of tourism and its transformation of physical and cultural reality.

From the above definition, it is clear that the use of "markers" is a necessary ingredient of trail definition, and these can, and do, take the form of physical signposting, information boards, viewpoint markers and other articulations of the landscape. The tourist trail therefore forms a narrative in which events and sites are thematically configured, but also one in which the configuration may be given direct physical expression. It is therefore relevant to consider the discourse of trail production as another example of the way in which tourism brings into being particular geographies.

Dartington Amenity Research Trust (1978) classified trails thematically into farm, nature, forest, town and ancient monument. The Scottish Tourist Board have used "categories" of attraction: people, town, architectural/archaeological and heritage (Silbergh *et al.* 1994). However, in addressing a perceived need to rationalize the process of trail development, in order to avoid duplication of themes and encourage integration at different geographical scales, Silbergh *et al.* propose a taxonomy of themed trails, using first a geographical scale (national, regional, local and site), then transport mode (car, bicycle, horseback, foot and public transport) and finally, ways in which the trail is marked or organized (signposted, printed literature, guided and self-guided). This is resonant of Lew's (1987) attempts to arrive at a taxonomy for tourist attractions. With the aid of their taxonomy Silbergh *et al.* hope that "a clear and ordered picture is projected to the visitor" (1994: 126). But following Fladmark's (1993) more imperious ideas for countryside interpretation, they propose that, nationally, themed trails might be planned to fit together like a set of Russian Dolls!

Conclusion

This chapter has examined the role of tourism in the geographical differentiation of the globe into spaces described as hyper-real, liminal and libidinal. This has been more of a conceptual discussion than an empirical one, for it is in the domain of theory that initial development is most urgently required. Tourism continues to be under-represented as a topic of geographical enquiry, but, when undertaken, has tended to approach the tourist use of space in a relatively unproblematic way—as an outcome of traditional spatial determinants such as the deterrence of distance.

In contrast, the approach advanced here has highlighted the active role of tourism in creating local geographies, semiotically, through the medium of place representation. Tourism, I have argued, differentiates space in a ceaseless attempt to attract and keep its market share. In the face of growing global cultural homogenization, local tourist agencies strive to assert their spatial distinctiveness and cultural particularities in a bid to market each place as an attractive tourist destination. This is achieved both intentionally, by the use of advertisements, brochures, press releases, travel agent promotions and education, and the like, but also unintentionally, as an effect of autonomous events such as the locational shooting of a film or television drama.

It has only been possible to address four such genres of place representation, but in each, attention was drawn to the ways in which considerations about the genre's production could be seen to influence its presentation. Each, it has been argued, produces as well as reflects the local geographies. Tourism is important to geographers, therefore, not simply as a result of its acknowledged global economic importance, but more significantly for its semiotic realization of the spaces that are nurtured in the tourist's imagination.

References

Adams, P. (1962) *Travelers and Travel Liars 1660–1800*, New York: Dover.

Bell, B. (ed.) (1995) *Scotland*, updated by Marcus Brooke, Hong Kong: APA Publications.

Boniface, P. and Fowler, P. J. (1993) *Heritage and Tourism in the Global Village*, London: Routledge.

Boyle, M. and Hughes, G. (1995) "The politics of urban entrepreneurialism in Glasgow," *Geoforum*, 25, 4: 453–70.

Britton, R. A. (1979) "The image of the Third World in tourism marketing," *Annals of Tourism Research*, 6: 318–28.

Cohen, E. (1973) "Nomads from affluence: notes on the phenomenon of drifter-tourism," *International Journal of Comparative Sociology*, 14, 2: 89–103.

——(1974) "Who is a tourist? A conceptual clarification," *Sociological Review*, 22: 527–55.

——(1979) "A phenomenology of tourist experiences," *Sociology*, 13: 179–201.

Coppock, J. T. (1982) "Geographical contributions to the study of leisure," *Leisure Studies*, 1, 1: 1–27.

Crang, M. (1994) "On the heritage trail: maps of and journeys to Olde Englande," *Environment and Planning D: Society and Space*, 12: 341–55.

Crick, M. (1989) "Representations of international tourism in the social sciences: sun, sex, sights, savings, and servility," *Annual Review of Anthropology*, 18: 307–44.

Dartington Amenity Research Trust (1978) *Self Guided Trails*, CP 110, Cheltenham: Countryside Commission.

Deller, H. and Stoelting, P. (1990) "Maps in travel guidebooks," *Cartomania*, Pelham MA: Association of Map Memorabilia Collectors.

Dilley, R. S. (1986) "Tourist brochures and tourist images," *The Canadian Geographer*, 30, 1: 59–65.

English Tourist Board (1987) *Effective Design and Print*, London: English Tourist Board.

Finn, A. (1988) "Print ad recognition readership scores: an information processing perspective," *Journal of Marketing Research*, 25, May: 168–77.

Fladmark, M. (1993) "Discovering the personality of a region: strategic interpretation in Scotland," in M. Fladmark (ed.) *Heritage: Conservation, Interpretation and Enterprise*, London: Donhead, 125–40.

Goss, J. D. (1993) "Placing the market and marketing place: tourist advertising of the Hawaiian Islands, 1972–92," *Environment and Planning D: Society and Space*, 11: 663–88.

Holbrook, M. and Lehmann, D. (1980) "Form versus content in predicting starch scores," *Journal of Advertising Research*, 20, 4: 53–62.

Hughes, G. (1992) "Tourism and the geographical imagination," *Leisure Studies*, 11: 31–42.

Jackson, P. (1989) *Maps of Meaning: An Introduction to Cultural Geography*, London: Unwin Hyman.

Lash, S. and Urry, J. (1994) *Economies of Signs and Space*, London: Sage.

Lefebvre, H. (1976) *The Survival of Capitalism*, London: Allen & Unwin.

Lew, A. A. (1987) "A framework of tourist attraction research," *Annals of Tourism Research*, 14: 553–75.

Lumsdon, L. (1992) *Marketing for Tourism: Case Study Assignments*, London: Macmillan.

MacCannell, D. (1973) "Staged authenticity: arrangements of social space in tourist settings," *American Journal of Sociology*, 79: 589–603.

——(1976) *The Tourist: A New Theory of the Leisure Class*. New York: Schoken.

Mansfeld, Y. (1990) "Spatial patterns of international tourist flows: towards a theoretical framework," *Progress in Human Geography*, 14: 372–90.

Motes, W., Hilton, C. and Fielden, W. (1992) "Language, sentence and structural variations in print advertising," *Journal of Advertising Research*, 32, 3: 63–76.

Ogilvy, D. (1983) *Ogilvy on Advertising*, London: Pan.

Otnes, H. M. (1978) *Index to Early Twentieth Century City Plans Appearing in Guidebooks: Baedeker, Muirhead-Blue Guides, Murray, I.J.G.R., etc., plus Selected Other Works to Provide Worldwide Coverage of over 2,000 Plans to over 1,200 Communities Found in 74 Guidebooks*, occasional paper no. 4, London: Western Association of Map Libraries.

Ousby, I. (1990) *The Englishman's England: Taste, Travel and the Rise of Tourism*, Cambridge: Cambridge University Press.

Poster, M. (ed.) (1989) *Jean Baudrillard: Selected Writings*, Cambridge: Polity Press/Blackwell.

Reimer, G. D. (1990) "Packaging dreams: Canadian tour operators at work," *Annals of Tourism Research*, 17: 501–12.

Riley, C. and Spackman, N. (1978) "Research to aid the design of cost effective holiday brochures," report on the ESOMAR Congress, Bristol.

Rossiter, J. (1981a) "Predicting starch scores," *Journal of Advertising Research*, 21, 5: 63–8.

——(1981b) "Visual imagery: applications to advertising," in A. Mitchell (ed.) *Advances in Consumer Research*, Pittsburgh PA: Association For Consumer Research, 9.

——(1988) "The increase in magazine ad readership," *Journal of Advertising Research*, 28: 35–9.

Selwyn, T. (1990) "Tourist brochures as post-modern myths," *Problems of Tourism/Problemy Turzstyki*, xiii, 3/4: 13–26.

——(1996) *The Tourist Image: Myth and Myth Making in Tourism*, Chichester: Wiley.

Shaw, G. and Williams, A. (1994) *Critical Issues in Tourism: A Geographical Perspective*, Oxford: Blackwell.

Shields, R. (1991) *Places on the Margin: Alternative Geographies of Modernity*, London: Routledge.

Short, J. R., Benton, L. M., Luce, W. B. and Walton, J. (1993) "Reconstructing the image of an industrial city," *Annals of the Association of American Geographers*, 83, 2: 207–24.

Shortridge, J. R. (1984) "The emergence of 'Middle West' as an American regional label," *Annals of the Association of American Geographers*, 74: 209–20.

Silbergh, D., Fladmark, M., Henry, G. and Young, M. (1994) "A strategy for theme trails," in J. M. Fladmark (ed.) *Cultural Tourism*, London: Donhead, 123–46.

Starch, D. (1966) *Measuring Advertising Readership and Results*, New York: McGraw-Hill.

Turner, V. (1974) *Dramas, Fields and Metaphors*, Ithaca NY: Cornell University Press.

Urry, J. (1990) *The Tourist Gaze: Leisure and Travel in Contemporary Societies*, London: Sage Publications.

Valiente, R. (1973) "Mechanical correlates of ad recognition," *Journal of Advertising Research*, 13, 3: 13–18.

Wicks, B. E. and Schuett, M. A. (1991) "Examining the role of tourism promotion through the use of brochures," *Tourism Management*, 24, December: 301–12.

Wordsworth, W. (1906) *Guide Through the District of the Lakes in the North of England*. 5th edn. London: Henry Frowde.

2

CYBERTOURISM AND THE PHANTASMAGORIA OF PLACE

Chris Rojek

The structure and consequences of cybercultures has become a major theme in recent academic literature.[1] The expansion of the Internet and the development of virtual places is viewed as heralding a major transformation in our understanding of the central categories of everyday life. To date, most of the speculation in this field has focused on the body and the information highway. For example, Haraway's (1991) work on cyborgs points to the incorporation of mechanical technologies into the skin, blood and fiber of the human body. Already, the installation of a pacemaker is regarded as a routine operation.

The merging of organism and microchip technology is evident in other areas as well. According to Wilson (1995), prosthetic technology and replacement surgery is undergoing a massive expansion. Within twenty-five years, it is predicted, the replacement of malfunctioning body parts with computerized microchip technologies and body parts from donors, to say nothing of cosmetic surgical improvements to the appearance of the body, will be standard medical practice, thus altering our current conceptions of aging and the life-cycle.

As for the effect of the information highway on local culture, Poster (1995) contends that the new technology will enrich consumer culture and expand choice. Rheingold (1994) speaks of the growth of "virtual communities" which are already emerging in cyberspace and which correct the decay of the public fora for democracy in the outside world. Plant (1993) combines the interest in body and space by arguing that cyberspace is compatible with swapping identities and liberating elements within us that are repressed in face-to-face interaction.

Not everyone has presented the new communications technology in meliorist terms. For example, Robins (1995) and McGuigan (1996: 182–4) have each poured cold water on the optimistic slant of much of the cyberculture literature. They acknowledge that cyberspace offers new potentials for human interaction and cultural formation. However, they stress that this potential also opens up the opportunity for the development of new types of

33

exploitation and control. What is surprising is that the subject of tourism is more or less absent from these discussions.

Conventionally speaking, we have understood tourism to be based in the movement of bodies between physical spaces. For example, the popular package tour business founded by Thomas Cook in England in the 1860s directly appealed to the yearning of Victorians for a break from domestic and work routines (Pemble 1987). It involved the regimentation and transportation of tourists through advertised scenic routes.

Victorian culture operated with an extremely sentimentalized vision of "home." Domestic space was identified as the foundry of character and the private refuge from the amorphous and turbulent "exterior" of society (Smiles 1859; Kellogg 1888). Mothers were stereotyped as the "angels of the household" and fathers as "providers and protectors." Leaving home and the home country was commonly seen as both an adventure and a hazard. This is why Cook went to such lengths to advertise and organize his tours as guided events. Cook's "popular holidays" offered the tourist the frisson of risk. Crucially, it was a controlled risk. The "Cookite" could travel abroad secure in the knowledge that professional tour guides had prearranged routes, hotels, foods, excursions and travel tickets.

Tourism today is far more varied and complex than in Cook's day. The package tour remains a staple feature of the industry, but it is supplemented by adventure holidays, semi-independent travel arrangements and a vast range of textually based, self-managed tours. However, the desire to have a break from the monotony and routine of domestic and work space remains a common motive behind tourist choice.

What is challenging about cyberspace is that it suggests that the notion of a break or rupture has been incorporated into the flow of everyday life. Domestic and work activities are punctuated with escape experiences and mind-voyaging through encounters in cyberspace. Our conventional categories of ordering space and classifying difference cease to be tenable when virtual worlds may now be created and experienced by travelers without physical relocation. As a result, the notion of physical and cultural space concentrated in activities related to escape and relaxation through tourism is now problematized. The core dichotomy which conventionally organized tourism experiences is the distinction between "home" and "abroad." Yet cybertravel has left this distinction in tatters, forcing us to rethink tourism's meaning within the context of contemporary society.

Tourism and the phantasmagoria of place

While the issues raised by cyberspace are not entirely new, it might be helpful to review them once again in order to sketch relevant features of the background discussion. The question of the limits of escape and relaxation experience in tourist sites has been widely debated (Lefebvre 1976, 1991; Urry

1990; MacCannell 1992; Craik 1994). An important theme in this literature is the nullification of the aura of tourist space by the growth of tourist infrastructure. Tourist space attracts tourist flows. It is a problem which is as old as the first recorded travel books such as Marco Polo's twelfth-century *Divisament dou Monde*, and *The Travels of Sir John Maundeville*, which began to circulate in Europe between 1356 and 1366.[2] As Pemble (1987: 170) notes, the elite tourist cultures of the late nineteenth century objected strongly to what the novelist George Gissing, called "the Cook's Tour type" muscling in on their retreats in the Mediterranean coast and the Holy Land.

The growth in holiday companies and tourist guides has a multiplier effect on tourist flows. The guides are associated with the expansion of what one might call the "Saint Thomas Effect." That is, the script describing the tourist sight instils a desire in readers to see the site for themselves. The effect of multiplying tourist flows is to alter the character of the destination. As Lefebvre (1991) puts it:

> Tourism and leisure become major areas of investment and profitability, adding their weight to the construction sector, to property speculation, to generalized urbanization. . . . No sooner does the Mediterranean coast become a space offering leisure activities to industrial Europe than industry arrives there; but nostalgia for towns dedicated to leisure, spreads out in the sunshine, continues to haunt the urbanite of the super-industrialized regions.
>
> (Lefebvre 1991: 353)

For Lefebvre then, the expansion of the tourist and leisure industries involves the commodification of space. The crucial irony here is that tourist space which is nominally intended for free expression, choice and self-determination is subject to rationalization and policing. The delineation of tourist space brings industry and commerce in its wake. Thus the quality of escape and the sense of sustainable difference from the rest of life is potentially damaged. However, Lefebvre also suggests that the longing for escape and relaxation outlasts the repeated disappointment and anti-climax of visiting over-exploited tourist sites.

The confinement and regimentation of urban-industrial existence is intolerable without a contrasting zone of time and space which is devoted to escape and relaxation. The specter of a world without escape is intrinsically alienating, since it suggests the attenuation of contrast and the decline of the mental and physical stimulation that contrast brings. In conditions in which home and work, town and country are reduced to a monotonous landscape, the vitality of the individual drains away.

The *coup de grace* of Lefebvre's discussion is that, unless controls on the commodification of tourist and leisure space are introduced, the costs of escape will eventually outweigh the benefits of staying at home. Escape and relaxation destinations will cease to deliver the experience of escape and relaxation. The result will be an appreciable increase in the alienation and despair of city-dwellers. Although Lefebvre does not spell out the social consequences in detail, it is clear that he has in mind mental unrest, illness, brutalization, rioting and the development of general asocial attitudes and behavior.

Benjamin (1976), working from a different line of analysis, reaches the same conclusion. His concept of "phantasmagoria" suggests the annihilation of stable meanings in culture and the convergence of public and private space. Benjamin found the kernel of the concept in the work of Marx, especially Marx's discussion of commodity fetishism. For Marx, capitalism generated a variety of commodities that offer satisfactions which are fundamentally illusory in character. What capitalism promises is an infinity of fulfilment based in the endless consumption and updating of goods. What it delivers is a limited supply of partial satisfaction and the inexorable replacement of the natural world with the artificial, synthetic, tumultuous commodity world.

Benjamin's Arcades project was dedicated to revealing the psychology of commodity culture.[3] It focused on retailing and merchandizing practices in the Paris of the Second Empire. Although Benjamin writes about Paris, his real subject is modernity and the transformations it exacts in terms of identity, association, practice and space, especially through generalized commodity consumption.

The redevelopment of Paris by Baron Haussmann after 1859 revolutionized the physiognomy of the city. Old, uneven vistas were replaced with regularized terraces of apartments and offices. The warren of medieval streets made way for straight, wide boulevards. The image of Paris was remade in the pure, smooth surfaces of the machine.[4] Haussmann's ambitious program of redevelopment uprooted many of the familiar reference points of old Paris. In particular, it threatened one of the principal settings for leisure, strolling, browsing and daydreaming: the Arcades.

The original arcades were established between 1822 and 1837. New industrial entrepreneurs required display areas to exhibit their goods, and the Arcades were designed to fulfil this function. As Benjamin (1976) points out, the Arcades—and Arcades culture—were an important source of attraction for foreigners. The display areas used a combination of glass, marble-floored passages, wrought iron and gas lighting to create an impression of allure and seduction. The retail philosophy was to make the setting of consumption seductive in order to enhance the appeal of the commodities and to attract customers.

Behind this was the desire to make the shopping experience appear to be leisurely and fulfilling so as to contrast with the regimentation and constraints of working life. Benjamin (1976: 158) quotes a contemporary illustrated guide to Paris which describes the Arcades as "a city, indeed, a world in miniature." For Benjamin, the Arcades created the illusion of inexhaustible variety and the satisfaction of all imaginable wants. They were tourist retreats within the city where the imaginative and fantasy possibilities created by modern manufacture were realized and explored. Williams (1982) and Saisselin (1985) have described them as conjuring "dreamworlds" of consumption. The Arcades created exotic zones of simulated escape and automated seduction in the heart of the urban-industrial complex.

The development of department stores continued and reinforced this process.[5] But crucially, the department store was a massified, centralized shopping experience which swept away the intimacy, quirkiness and secret delights of the Arcades. The department store offered consumption in a bureaucratized form, so that little of the charm and quirkiness of the Arcades remained.

For Benjamin, commodification was at the heart of the construction and stimulation of the urban phantasmagoria. Commodification constructs a metropolitan environment which revolves around artifice. Calculated illusion, metronomic spectacle and automated seduction masks the process of buying and selling. In Marx, the exploitative side of commodification is stressed. His analysis depicts the consumer becoming more and more enmeshed in the logic of capitalist accumulation. The ideology of freedom and choice perpetuated in consumer culture is a central mechanism in reproducing false consciousness.

The consumer forgets the exploitation at the heart of the capitalist exchange relationship. Instead, through an array of advertising and marketing devices, the consumer learns to associate consumption with pleasure and fulfilment. In Marx's view, this was the epitome of false consciousness because consumers were trained to celebrate the processes which contributed to their enslavement.

Benjamin retains Marx's emphasis upon exploitation and false consciousness. But he gives it an important new twist. He maintains that the phantasmagoria of consumer culture reveal the repressed, utopian wishes of the masses. Advertising and illusion do not simply deceive the consumer, they register an authentic trace of popular hidden wishes. In this way, Benjamin claims a utopian moment in ordinary consumer relations. His reading of commodification suggests both mass amnesia and mass awakening.

This is why the Arcades are a crucial test for his theory of modern consumer psychology. The individuality, elegance and variety of the Arcades has been smashed by the standardized consumption experience of mass retailing and department-store services ushered in by Haussmann's urban redevelopment program. Yet the masses remain surrounded by the broken shell of individualized and luxury retail outlets, which provide a constant reminder of the true capacity of modern technology and organization to satisfy desires and wishes. Moreover, this shell acts as a reminder of the general limits imposed upon human relations by the system of capitalist production. In Benjamin's words, the phantasmagoria of the capitalist system of production correspond to images in the collective consciousness in which the new and old are intermingled. These images are ideals, and in them the collective seeks not only to transfigure, but also to transcend, the immaturity of the social product and the deficiencies of the social order of production. In these ideals there also emerges a vigorous aspiration to break with what is out-dated—which means, however, with the most recent past. These tendencies turn the fantasy, which gains its initial stimulus from the new, back upon the primal past. In the dream in which every epoch sees in images the epoch which is to succeed it, the latter appears coupled with elements of prehistory— that is to say of a classless society. The experiences of this society, which have their store-place in the collective unconscious, interact with the new to give birth to the utopias which leave their traces in a thousand configurations of life, from permanent buildings to ephemeral fashions.

(Benjamin 1976: 159)

Tourism is a key example of phantasmagoric culture because it typically involves the clash between old and new, and is based in experience of contrast and escape. In traveling, we reveal the limits of our own personal and cultural worldviews, as well as encountering customs, habits and values which differ from our own. Tourism shows us new ways of organizing personality and life space, and exposes the socially constructed character of our beliefs and values. It also carries a critical potential in contrasting the gnomic character of our routinized existence with the simultaneous worlds which are in reach merely by taking a car ride or buying a plane ticket.

However, Benjamin's work on the phantasmagoria of place also provides a rationale for believing that the traditional appeal of tourism as a way of gaining a break from the routine and monotony of everyday life is rapidly fading. He argued (1976: 167–8) that in the time of the Second Empire, the bourgeois interior gradually crystallized into "the universe for the private citizen. In it he assembled the distant in space and in time. His drawing-room was a box in the world theatre." Confronted with the swirling mass of anonymous humanity in the exterior world of city streets and squares, the individual repaired to the domestic interior wherein he had the liberty to develop and leave traces of his real self. Freed from the obligation of the work contract to devote time to the service of adding exchange value, the individual's leisure and domestic space became the place where one can really "be oneself." The bourgeois interior is fashioned as a place of refuge and a site of dreams. It is crammed from the floor to the rafters with personalized souvenirs and furniture chosen to signify status.

Benjamin speaks of "the knick-knacks on the shelves, the antimacassars on the arm-chairs, the filmy curtains on the windows, the screen before the fireplace" (quoted in Reed 1996: 10). The motivation behind this accumulation of a storehouse of commodities is to externalize one's own sense of personal worth and to signify one's distinctiveness to the rest of the world. Yet Benjamin fully realized that the personality which was displayed with such meticulous attention to detail was, in fact, mass-reproduced. However different each bourgeois interior may have been to other domestic living spaces in the nineteenth century, it was closer to other bourgeois interiors than to anything else. Therefore the sense of personal worth and distinctiveness that leisure and personal living space were designed to convey is rejected by Benjamin as a delusion.

Nor was this sense of the delusional character of the bourgeois interior confined to Benjamin. The novelist J. K. Huysmans (1884) captured it in artistic form in his novel *A Rebours,* which deals with the baroque imagination of the sickly Comte Des Essientes and his decadent attempts to simulate passion and escape in the reconstructed spaces of his decaying private mansion. As if to rub it in that the bourgeois perception of inhabiting "a box in the world theatre" was in fact a delusion, Benjamin pointed to the architectural developments of the early twentieth century which were beginning to erase the line between public and private space.

The anti-decorative, rational architectural designs aggressively pursued by Adolf Loos and Le Corbusier were transforming the notion of personalized space. These architects associated purity and a lack of local references with progress. The new steel and glass structures which they erected in public space created a built environment in which, as Benjamin (1977: 218) remarked, "it is difficult to leave a trace." These design values were carried over into domestic space. The domestic interior was now conceived as a machine to assist human functions. Decorative style was associated with waste and distraction. The new emphasis was on facilitating concentration and rational living. Loos (1908), in his article "Ornament and crime," went so far as to argue that the more refined a species is, the less inclined it is to decorate (quoted in Anger 1996: 130).

Benjamin refers to these designers as symptomatic of a trend to converge private and public space. The effect of this movement is to destroy the divisions which make travel worthwhile. Why leave the domestic interior if one's destination is a replica of it? This is the question that Lefebvre (1976, 1991) posed in the course of his discussion of the commodification of tourist space.

Benjamin's own analysis of the convergence between private and public space points in the same direction. But to it he adds a significant new question: Why leave the domestic interior if it offers a richer escape environment than dedicated leisure and tourist space? Postmodern architectural style is a reaction to what is now regarded as the over-rationalistic, over-mechanical design values of Loos and Le Corbusier (Jencks 1984). It celebrates the exuberance of conjoining formally distinct styles and the incorporation of local references, fantasy and irrationalism into a single building design. Yet this departure from the absolutism of modernist architecture has not put an end to the convergence of public and private design values. It is just that the dynamics of this convergence have changed.

The central metaphor of modernist design was the machine; in the post-modern period it is the communications network. Vidler (1996: 178) writes of the emergence of "homes for cyborgs" in which "private space is revealed as infinitely public, private rituals publicized to their subjects and those in turn connected to the public matrix." However, cyberspace is different in as much as these connections are made from within the domestic interior. Virtual reality technology allows the individual to explore a computer-generated place without leaving the home. Following Rheingold (1994), Poster concentrates upon the positive effect of the new technology. He writes:

> A participant may "walk" through a house that is being designed to get a feel for it before it is built. Or s/he may "walk" through a "museum" or "city" whose paintings and streets are computer-generated but the position of the individual is relative to their actual movement, not to a predetermined computer programme or "movie."
> (Poster 1995: 85)

Of course, Poster is only speculating, and the use of scare quotes suggests a lack of confidence that the social forms which will emerge from this technology will be qualitatively different from forms associated with bodily movement and travel through physical space. Nonetheless, the development of virtual reality logically implies the opening-up of new social spaces in contemporary culture.

One way in which these spaces may be increasingly used is through the practice of cybertourism. That is, the exploration of computer-generated distant places from the safety and comfort of one's home. If Poster (1995) is to be believed, the development of cyberspace has the potential to genuinely realize Benjamin's proposition that the interior offers a box in the world's theater. Moreover, this time the escape route is not only confined to the bourgeois leisure class but is open to the masses. The new virtual technology permits us to roam throughout the world and accumulate tourist experience without ever boarding a mechanism of physical transport. Cybertourism appears to offer a quantum leap in our potential to visit distant sites and accumulate direct experience of different cultures and cultural objects.

Yet on balance, the triumphalist tone in the cyberspace lobby should be resisted. Three points must be made. First and foremost, the innovation of the technology is occurring in the social context of structured inequality. Academics, media people and other members of the service class may have the surplus cash and time to buy the digital equipment necessary to participate in the newly emerging virtual worlds, but this is not the case with the population at large. The excluded consist of those without full-time paid employment. That is, people who are either unemployed and reliant on state benefit or are working in temporary, part-time work where the conditions of paid labor and the security of return are uncertain. Their access to a telephone, let alone digital technology capable of generating virtual reality, is strictly limited (Thomas 1995).

The second point is that the quality of the escape experience offered by virtual reality is not equivalent to the tourist experience of physical movement through space. Cybertourism is really an extension of the ordinary daydreaming activity explored by Bachelard (1970) in his work on revery. The sense of authenticity sustained in computer-generated space may be of a high order, but it is qualitatively different to actually being in physical space. Most obviously the experience takes place inside the head. Whatever sensory simulation is introduced in order to enhance the effect of real experience, it is unlikely to outweigh the paramount sense of having a travel experience which is concentrated in one's mental life.

Poster's (1995: 86) account of multi user domains (MUDs) on the Internet implies that interactive participation may eventually be compatible with computer-generated virtual reality. But for the moment the technology has not been realized. Despite the expansion and release of imagination and identity experimentation claimed for the Internet and virtual technologies, the experience remains essentially solipsistic and confined to the head rather than the body.

The third point is that this virtual technology does not provide a genuine or convincing world of escape. Our real life hopes, desires, worries and fears are not left behind when we enter cyberspace. Rather, they are translated into another key of consciousness and interaction. Through cybertourism we may "see" the Statue of Liberty and walk through the corridors of the Prado and the British Museum, but that does not stifle the desire to see these places for ourselves. The "Saint Thomas Effect" alluded to above is perhaps more important in explaining tourist motivation than the advocates of cyberspace and cybertourism allow.

None of this is to discount the importance of cyberspace in redefining some aspects of our orientation to external space, or to reject the idea that it diversifies people's escape experience. Our travels in cybertourism are far from being fated to be of negligible or trivial importance. Through them, we gain impressions and memories that we retain and which act as factors in relating to others in the real world. But cyberspace does not replace the real world. In assessing the experiences that we gain through cybertourism we are driven back to real-world encounters for a depth perspective.

Schutz's (1967) idea of a binding "paramount reality" which acts as the context for all forms of social encounter is not negated by the rise of cyberspace. On the contrary, the punctuation of routine experience by cyberspace may be said to strengthen our dependence on lived culture with others and documentary records as the foundations of the life-world. For when we are surrounded by simulated escape experience we reinforce our requirement to hang onto something familiar and dependable. A world in which cyberspace replaced the phenomenal categories of paramount reality would be as alienating and inhuman as a totally commodified society in which resistance ceases to be possible. Marcuse (1964) realized this after his depressing discussion of "one dimensional society" which is why his later work on liberation and aesthetics provided a much greater place for the capacities of interpretation and negotiation in human actors.

Commodified escape routes

The urgent and fundamental questions that we face in the sociology of tourism are not posed by the imminent growth of cyberspace into our leisure time and space. My assessment of the practical potentials offered by cyberspace and cybertourism has been designed to put these new developments in perspective. The critical perspective against which these potentials must be measured is the much more deeply rooted commodification of tourist space. As MacCannell (1992: 176) notes, the crucial questions derive from the paradox that the consequence of mass tourism is to homogenize tourist space through the processes of commodification, while its ideology retains an emphasis upon escape and difference as the hallmarks of authentic tourist experience.

Concrete increasingly covers the paradisical islands, sandy beaches and mountain retreats that our culture has traditionally identified as the embodiment of "nature" and a principal object of tourist escape experience. To repeat Lefebvre's (1976, 1991) point, this can only be resisted effectively by planning regulations. Most government and tourist bodies are reluctant to contemplate authorizing restrictions on the tourist infrastructure because to do so invites the risk of depressing the tourist economy. Yet at the national and international levels, regulation does exist in the form of protective legislation for areas of outstanding beauty and world heritage sites.

As one example, UNESCO operates at the general international level to establish "objective" cultural parameters for developments within nation-states and regions (Tomlinson 1991: 70–4). Plans to extend road networks through areas of outstanding natural beauty or to threaten world heritage sites by economic exploitation now involve entrepreneurs in facing rather more than the local city or federal court. For example, in 1993 the Indian Supreme Court ordered that 200 factories near the Taj Mahal must close because industrial pollution from the factories was threatening the building.[6] The ruling represented a success for local, national and international environmental and heritage defense pressure groups who brought influence to bear upon the Indian government.

Examples like this suggest the existence of a new level in policing tourist space. But, while conservation has done much to protect tourist environments, it has the unintended consequence of sequestering specific tourist spaces from the ordinary flow of life and cannibalizing the lives of the people who are situated there. Conservation has been criticized for instilling a lifeless quality into tourist space by devouring its living relationship with the rest of society. MacCannell (1992: 17–73) refers to the development of "ex-primitive" populations in the main third-world tourist routes. By this term he means individuals who are required to exhibit the characteristics of primitive life solely for tourist consumption.

The phenomenon points to an interesting moral dilemma. The thrust of industrialization is to obliterate the life of agrarian and hunting tribes by co-opting tribal members into the processes of nation-building and citizenship. Thus the preservation of the life of hill tribes and forest peoples may be judged to be a contribution of the Western tourist industry and morally good. On the other hand tourism forces a regime of lifelong performativity upon these members, which is alienating and demeaning. The tribal members are often college-educated with homes in the city, yet they are forced to play the part of primitives. Hence tourism participates in the preservation of historically redundant ways of life, but only by confining them to exhibition value. The idea that ex-primitives might really be foraging and hunting, without recourse to Western medicines and beyond the control of Western law, would destroy the principle of predictability upon which the tourist experience is based.

Although MacCannell's (1992) discussion is predicated in the experience of the pre-industrial ex-primitives rather than the industrialized tourists, it is clear that a parallel development is occurring in the post-Fordist economies of the West. MacCannell himself (1992: 172–80) refers to the repackaging of an entire Chinese community in the town of Locke in California. The Asian buyer "purchased" an "authentic" working Chinese community in the Sacramento valley and marketed it as a tourist attraction. Similar developments can be found in Torrance, California, which boasts an "alpine village" complete with a reconstruction of the buildings and streets of a typical Bavarian settlement and a full-time complement of Bavarian peasants dressed in authentic costume; and also in Helen, Georgia, which boasts a reconstruction of a Swiss chalet-style village.

More generally, bankrupt coal mines, ironworks and cotton mills are being re-exploited as tourist attractions. Communities of what might be called "ex-industrials" are employed to operate the machinery and play the parts of authentic industrials. The paradox here is very strong. Tourists, who are basically involved in a leisure activity, pay to watch people who would otherwise be unemployed performing jobs that can no longer be sustained in the industrial sector. Of course, there are strong precedents for using performative labor as a tourist attraction in the heritage industry. A "The Way We Were" exhibition at Wigan Pier employs a company of actors dressed in period costume to enact life at the turn of the century for day-trippers and other holidaymakers; similar dramaturgical devices are employed in the North of England Open Air museum at Beamish in County Durham, an identikit heritage town created from reconstructed bits and pieces from England's urban-industrial past; the "Great American Theme Park" in the Midlands; the settler community reconstruction at the Plymouth plantation in New England; and the presentations of traditional life in "typical" English, Italian, French and German villages in Busch Gardens, Williamsburg.

All of these developments present a cannibalized version of history which is designed to maximize tourist flows. The accretion of the same techniques in the industrial sector, at a time when the mass manual workbase is contracting, is revealing. It suggests a nostalgia for the recent industrial past in which people had real jobs and worked in communities as opposed to the privatized, short-term and uncertain job market of the present. The recommodification of work activity is also revealing. We know from the Marxist tradition that workers under capitalism experienced work as alienated labor. Typically, they had no security and no control of the work process including the distribution and sale of the commodities which they created. Some of these workers are now reallocated into performative industrial tasks in which their commodified value is redefined as exhibition fodder for tourists.

Of course, there are deep disputes about the cultural meaning of these developments. For example, Hewison (1987) regards the development of ex-primitives and ex-industrials as producing a gloss on our relation with history. "The heritage industry," he writes (1987: 10), "draws a screen between ourselves and our true past. . . . Hypnotized by images of the past, we risk losing all capacity for creative change." Others, like Wright (1985) and Samuel (1994) question the idea of "our true past." They argue that the populist heritage industry provides a counter to elitist and condescending constructions of "our" history, while MacCannell (1992) regards the creation of strata of perfomative cultures designed to simulate the pre-industrial and industrial world as one of the ultimate, corrosive delusions of modernity.

Conclusion

What emerges most powerfully from the literature is a strong tendency in the culture of tourism for escape activity to adopt a commodified form. To repeat: tourist space attracts tourist flows. The designation and management of escape enclaves typically results in the production of a tourist infrastructure consisting of road and rail networks, hotels, viewing and refreshment sites, souvenir shops and parking spaces, which curtails our experience of escaping from the pressures of urban-industrial culture. Other tourists become the biggest enemy of the tourist seeking escape. The tawdry collection of shops and refreshment areas on the northern edge of Niagara Falls, the guided bus tours which interrupt you as you stand in front of the Washington Memorial or the Statue of Liberty, and the camera-snapping, backpacked queues that throng Oxford Street and Picadilly Circus in London, serve to choke off the sense of leaving ordinary life behind that the tourist brochures promise the experience of travel will deliver. Of course, tourist cultures are not static. Feifer (1986) and Urry (1990) argue that already there are signs of a "post-tourist" response to the crowding-out of holiday destinations.

Post-tourism involves the cultivation of an ironic disposition to the tourist site. The post-tourist accepts that the site will be swarming with other tourists and treats this as part of the tourist experience. The Romantic ideal of being alone with the tourist object so that one can possess it fully is abandoned. Post-tourists have come to terms with the commodified world and do not hanker after pre-commodified experience. Perhaps the degree of rational reflexivity which Feifer and Urry attribute to the post-tourist is a little exaggerated. After all, escape and contrast are still the most prominent themes in the ideology of tourism. It is the irrational possibility of leaving the ordinary, over-familiar world behind which persuades millions of us to spend our savings on trips designed to "get away from it all." Joining a queue of others intent on achieving the same experience must tax the patience of even the most dedicated post-tourist.

Interestingly, the sheer volume of tourists moving from place to place in the world today means that tourist counter-cultures which grew up in the late 1960s and 1970s are themselves becoming more intensively commodified. These counter-cultures were typically very critical of established tourist routes and the package deals offered by tourist operators. They emphasized authenticity and spontaneity over the managed, plastic experience offered by established tourist chains.

One of the most successful examples of the tourist counter-culture is the *Lonely Planet* guidebook series. This celebrated the merits of self-help, independent travel, and value for money. The guides are aimed at the long-distance travel market. They encourage tourists to be well prepared and to enjoy the experience of different cultural values. The anxiety of being abroad which Thomas Cook exploited in the nineteenth century is replaced with a robust confidence in the ability of the tourist and the guidebook to solve any travel problems. The basic Lonely Planet travel philosophy is: "Don't worry about whether your trip will work out. Just go!" (Finlay 1993: 104).

The guides have been enormously successful. In 1993, Lonely Planet employed over fifty authors and fifty-four staff (editors, cartographers and designers) at their Melbourne headquarters. Satellite offices operated in London, Paris and Oakland, California. The organization publishes a free quarterly newsletter, *Planet Talk*, and has diversified into producing a range of "Lonely Planet" t-shirts. Each issue of the newsletter and each new edition of the guide carries an open invitation for readers to share their travel experience with the editors. This acts as a free information-gathering service which is used to update new Lonely Planet publications.[7]

The company encourages tourists to be respectful of the landscapes and cultures that they visit. A percentage of the income from each book is donated to the countries it covers. In 1993, US$100,000 was donated to causes such as famine relief in Africa, aid projects in India, agricultural projects in Central America, and Greenpeace's efforts to prevent French nuclear testing in the Pacific. The dominant motif used in the literature is of a family of Lonely Planet tourists bringing the positive, enriching merits of tourism with them. Yet the numbers of guides sold since the publication of the first guide in 1973, *Asia on The Cheap*, also mean that hundreds of thousands of people have trodden along the "off the beaten track" locations listed in the guide and stayed at the recommended "little known, private" hotels and guest rooms. As a result, the "alternative" destinations have lost the charm of being truly secret or off-the-beaten-track.

The pressure of tourist bodies upon tourist spaces is undeniable. It produces complaints from host communities who feel engulfed by a tidal wave of travelers, and also from other tourists who find that it is harder and harder to get away from it all. To some extent, the irrational desire to discover pre-commodified worlds becomes translated onto another plane. It is surely no accident that youth cultures today show such a strong interest in the occult, the supernatural and science fiction. Television programs like *The X-Files* and *Star Trek*, and films like *Independence Day* cater for the desire to leave the commodified world behind. Ironically of course, they only exist in commodi-fied units to be bought and sold in the entertainment industry. Yet they also register the real utopian desire for finding something which is different from the commodified phantasmagoria of Western capitalism. This is the same longing that the tourist industry exploits. It stems from the common feeling of being overwhelmed by a programmatized, bureaucratized, alienating world.

The irony is that the result of dreaming our dreams of escape is to extend the reach of this alienating world into our enclaves of escape and relaxation. The possibility of personal escape is contradicted by the realization of the mass yearning to leave the familiar everyday world behind. This is the tragedy, and also the condition, of mass tourism today.

Notes

1   This reflects a broader trend toward an interest in signification and the regula-tive/enabling principles of sign economies (advertising, television, film, pop music). Basically this interest coincided with the rise of cultural studies in the university system which started from the late 1970s.
2   Maundeville's book purported to be the memoirs of a fourteenth-century English knight. However, it is probable that the book was a composite of travelers' tales. The artifice, fabrication and calculated deception which many critics of modern travel and tourism have found in the tourist industry evidently has a much longer pedigree in the historical formation of tourist culture. Adams (1962) provides a fascinating survey of travel lies and myths perpetuated between 1660 and 1800.
3   Among other things he was literally interested in the disembeddedness that the continuously changing vista and character of metropolitan culture exerted on the psyche of the individual. Following Simmel, Benjamin interpreted modern exis-

tence as a succession of shocks to the system. While he did not go as far as Simmel in claiming that these shocks produced psychological pathologies (neurasthenia and the blasé attitude), he nevertheless identified the growth in anxiety and nervousness as a primary characteristic of the rise of modern societies.

4   The metaphor of the machine was much to the fore in Haussmann's mind. He saw the radial boulevards, terraces and squares of the new Paris as delivering the efficiency and reliability of the best in the machine age.

5   Wood's (1988) study of the show windows in department stores charts an exotic history of the staged tableaux that store chiefs mounted in order to attract publicity and customers. The use of live animals, rare plants and paid actors were all deployed in order to increase the attraction value of stores. This reinforced the sense of the metropolis as a phantasmagoric place, a place in which nothing and everything is real.

6   The order raised interesting questions of priorities in India's path to modernization. In particular, should the interests of foreign tourists bearing foreign currency be allowed to weigh over the interests of local industrialists intent on establishing an industrial home base which would supply jobs for thousands of workers? While certainly open to interpretation, the court order does suggest that developing economies may have few credible options with which to force industrial change when it threatens global heritage and travel interests.

7   Interestingly, this confirms one aspect of Ritzer's (1996) McDonaldization thesis, which claims that one of the key trends in modern consumer culture is to co-opt the consumer in the production of services. Ritzer uses the example of the McDonald's restaurant where the consumer is encouraged to dispose of the wrappings and cartons left over after eating the food. Lonely Planet incorporates feedback from consumers of its travel guides to improve the information included in future guides.

References

Adams, P. (1962) *Travelers and Travel Liars 1660–1800*, New York: Dover.

Anger, J. (1996) "Forgotten ties: the suppression of the decorative in German art and theory, 1900–1915," in C. Reed (ed.) *Not At Home: The Suppression of Domesticity in Modern Art and Architecture*, London: Thames and Hudson, 98–112.

Bachelard, G. (1970) *Le Droit de Rêver*, Paris: Presses Universitaires de France.

Benjamin, W. (1976) *Charles Baudelaire: A Lyric Poet in the Era of High Capitalism*, London: Verso.

——(1977) *Gesammelte Schriften*, vol. 2, Berlin: Suhrkamp.

Craik, J. (1994) *Resorting to Tourism*, Sydney: Allen and Unwin.

Feifer, M. (1986) *Tourism in History*, New York: Stein and Day.

Finlay, M. (1993) *India: Lonely Planet Guide*, Melbourne: Lonely Planet Publications.

Haraway, D. (1991) *Simians, Cyborgs and Women*, London: Free Association Books.

Hewison, R. (1987) *The Heritage Industry*, London: Methuen.

Huysmans, J. K. (1884) *A Rebours*, Harmondsworth: Penguin.

Jencks, C. (1984) *The Language of Postmodern Architecture*, London: Academy Press.

Kellogg, J. (1888) *Plain Facts for Old and Young*, Burlington IA: Burlington Press.

Lefebvre, H. (1976) *The Survival of Capitalism*, London: Allen and Unwin.

——(1991) *The Production of Space*, Oxford: Blackwell.

MacCannell, D. (1992) *Empty Meeting Grounds*, London: Routledge.

McGuigan, J. (1996) *Culture and the Public Sphere*, London: Routledge.

Marcuse, H. (1964) *One Dimensional Man*, London: Abacus.

Pemble, J. (1987) *The Mediterranean Passion*, Oxford: Oxford University Press.

Plant, S. (1993) "Beyond the screens: film, cyberpunk and cyberfeminism," *Variant*, 14: 12–17.

Poster, M. (1995) "Postmodern virtualities," *Body and Society*, 1: 3–4, 79–96.

Reed, C. (1996) (ed.) *Not At Home: The Suppression of Domesticity in Modern Art and Architecture*, London: Thames and Hudson.

Rheingold, H. (1994) *The Virtual Community: Finding Connection in a Computerised World*, London: Secker and Warburg.

Ritzer, G. (1996) *The McDonaldization of Society*, Thousand Oaks CA: Pine Forge Press.

Robins, K. (1995) "Cyberspace and the world we live in," *Body and Society*, 1: 3–4, 135–96.

Saisselin, R. (1985) *Bricabracomania: The Bourgeois and the Bibelot*, London: Thames and Hudson.

Samuel, R. (1994) *Theatres of Memory*, London: Routledge.

Schutz, A. (1967) *The Phenomenology of the Social World*, London: Heinemann.

Smiles, S. (1859) *Self Help*, Harmondsworth: Penguin.

Thomas, R. (1995) "Access and inequality," in N. Heap, R. Thomas, G. Einon, R. Mason and H. Mackay (eds) *Information, Technology and Society*, London: Sage, 131–51.

Tomlinson, J. (1991) *Cultural Imperialism: A Critical Introduction*, Baltimore MD: Johns Hopkins University Press.

Urry, J. (1990) *The Tourist Gaze: Leisure and Travel in Contemporary Societies*, London: Sage.

Vidler, A. (1996) "Homes for cyborgs," in C. Reed (ed.) *Not At Home: The Suppression of Domesticity in Modern Art and Architecture*, London: Thames and Hudson, 161–78.

Williams, R. (1982) *Dream Worlds: Mass Consumption in Late Nineteenth-Century France*, Berkeley CA: California University Press.

Wilson, R. (1995) "Cyber(body) parts: prosthetic consciousness," *Body and Society*, 1, 3–4: 239–60.

Woods, R. J. (1988) *Confessions of a PR Man*, New York: NAL Books.

Wright, P. (1985) *On Living in an Old Country*, London: Verso.

Part II

DESTINATIONS

3

LANDSCAPE RESOURCES, TOURISM AND LANDSCAPE CHANGE IN BALI, INDONESIA

Geoffrey Wall

Tourism is both an agent of preservation and change of landscapes and landscape elements. Many destination areas are noted for their special landscapes, and tourism provides an impetus and an economic rationale for preserving valued features. At the same time, the introduction of tourists into an area, plus the facilities to cater to their needs, inevitably results in landscape change. In some situations, as in many coastal and mountain resorts, the growth of tourism can result in the creation of novel landscapes which are so strongly influenced by tourism that it is reasonable to label them as tourist regions with tourism landscapes. Thus tourism is at the same time both a conservative and a radical force in landscape evolution.

The pace and scale of change associated with tourism vary considerably from place to place and, for any particular place, from time to time. Also, traditional, relict, modified and new landscape features may be juxtaposed in a kaleidoscope of land uses and visual forms. Furthermore, these features may have different meanings to different people. At a very basic level, for residents, the landscape may be one associated primarily with work and everyday living, whereas for visitors it may be a landscape of pleasure experienced in a brief sojourn. In some developing countries where the tourism economy is dominated by outside investors, the landscape may reflect differential access to power, with tourism being viewed as a form of neo-colonialism by many local people and as a means of development by an elite (Nash 1989).

Even where the same features are valued by both tourists and visitors, they may be valued for different reasons: perhaps as sites and places to be lived in and possessing profound personal, cultural and religious significance for local residents, as compared with sights to be viewed, passed by and, perhaps, captured on film by the traveler (Hull and Revell 1989). The differences in backgrounds and interests between locals and visitors suggest that it may be necessary to interpret the local landscape and its meanings to visitors, and a profession of interpreters has arisen to meet this need (Knudson *et al.* 1995). However, their task is not an easy one, for the landscape may tell many stories and its meanings may be contested by different groups.

This contribution examines the landscape of Bali as it has been and continues to be modified by tourism. The discussion will occur at a variety of scales and with regard to the meanings traditionally ascribed to the landscape by the Balinese as well as the ways in which the landscape is being modified by the forces of tourism, sometimes in congruence but often in conflict with Balinese traditions. But first it is necessary to place Bali and its tourism in the context of Indonesia and its development needs.

The Indonesian context

Indonesia is a country with great size and diversity. The Indonesian archipelago encompasses in excess of 13,500 islands and stretches a greater distance from east to west than North America (Wall 1997a). Indonesia is home to more than 220 million people representing more than 300 ethnic groups and speaking more than 250 different languages. While located entirely in the tropics, mountains in Irian Jaya have snow on their peaks. Moist regions in the west give way to drier and even arid areas in the east. The Asian and Australasian geological plates are juxtaposed in Indonesia, and this is an important reason for the great differences in flora and fauna between east and west, and a major reason for Indonesia possessing some of the greatest biodiversity among nations (Micheaux 1994).

Levels of development also vary markedly, generally declining in an easterly direction, and the sophistication of parts of Jakarta, the capital, contrasts heavily with the traditional lifestyles still found in parts of Kalimantan and Irian Jaya. Thus for a mix of reasons based in both physical and human phenomena, the landscapes of Indonesia are diverse. It is this diversity of both natural and cultural features that constitutes a rich resource for tourism.

Faced with declining oil prices, Indonesia has turned increasingly to tourism as a source of foreign exchange. Tourism has received increased prominence in each successive *repelita* (five-year plan) and the number of foreign visitors and their expenditures has risen substantially in recent years. The number of international visitors more than doubled between 1987 and 1991, with annual growth rates averaging in excess of 27 percent in this period. In 1987 there were 1,060,347 international visitors; the two-million mark was breached in 1990; 2,569,870 visitors came in 1991; and 3,064,161 arrived in 1992. In spite of the Gulf War, an 18 percent growth rate was achieved between 1990 and 1991, reflecting the success of Visit Indonesia Year, rising to 23 percent between 1991 and 1992 (Wall 1997a). In 1996, almost 4.5 million international visitors entered Indonesia (WTO 1997).

Much of this tourism has been mass tourism in "sea—sun—sand" settings, but cultural and other natural features have played strong supporting roles.

Tourism is now third behind textiles and wood products as a non-oil generator of foreign exchange, and is projected to rise to first place by the year 2000. Thus when viewed from a national perspective, the recent rapid rate of growth of international tourism can be viewed as a great success.

If the scale of analysis is changed, one finds that this level of success should be qualified in a number of ways. As with economic development as a whole, tourism in Indonesia is extremely unevenly distributed. Roughly a quarter of international travelers visit Jakarta, the capital, but many of these are primarily on business. Another quarter visit Bali, and a growing number of short-stay visitors are entering Indonesia from Batam in neighboring Singapore. While there are a substantial number of other established international tourist destinations, and also a growing domestic market, most areas of Indonesia have yet to be modified greatly by tourism, although this is beginning to change.

However, where tourists are concentrated, as in Jakarta where the landscape is dotted with high-rise hotels, and in Bali where tourism, along with agriculture, dominates the economy, its effects are becoming increasingly visible. The remainder of this chapter will concentrate on the case of Bali, Indonesia's premier tourism destination, and will discuss the creation of new landscapes in Bali which meld the traditional and the modern.

The natural endowment of Bali

The island of Bali is approximately 563,00 hectares in area, making it about the size of Prince Edward Island, Canada. It has a volcanic core with calderas, lava flows and a limited number of upland lakes. Forested uplands give way to well-watered terraced slopes where the *sawah* (irrigated rice fields) constitute a beautiful, manicured, highly productive, humanized landscape which is also a tourist attraction. The island is fringed by both white and black sand beaches, cliffs, reefs and mangroves, the last remaining large example of which is to be found in Benoa Bay in the south.

There are important regional variations in this landscape. The north is in a rain-shadow area and the north and east are generally drier than the south and west. Also, the white sands, which are preferred by tourists, are concentrated in the south of the island.

The cultural veneer

Bali has been called a Hindu Island in a Moslem sea. While the majority of Indonesians are officially devotees of Islam, more than 90 percent of residents of Bali are Hindu. For the Balinese, the lands and waters and the associated landscapes provide both physical and spiritual sustenance, for the landscape is infused with religious significance at a variety of scales.

The following brief introduction to Balinese cosmology is based primarily upon the work of Budihardjo (1986). The basic concept of *Tri Angga* divides everything into three components: *Nista* (low, impure, leg), *Madya* (middle, neutral, body) and *Utama* (high, pure, head). The physical world, natural life and time are all treated as if they are each composed of three parts, e.g., atmosphere—lithosphere—hydrosphere, mountain—land—sea, gods—humans—evil spirits, past—present—future. Balinese people are oriented toward the interior of the island rather than to the sea. They have not generally been a seafaring people, and although some ceremonies are associated with the sea and a number of important temples have coastal locations, the sea is generally associated with evil spirits. Rather, the Balinese look to Mount Agung, the highest point on the island (3,142 meters), where the Gods are presumed to abide.

To the Balinese, east and west are much less important directions than *kaja* and *kelod*, toward the holy mountain and away from the holy mountain. As will be seen, these directions play a fundamental role in the orientation of Balinese villages and family compounds. The threefold division into mountain, productive land and sea can be compared to the human form of head, body and legs. The high point is venerated, and in Bali the head is respected and one should not touch the head of another or raise one's position above that of the body at a cremation.

The network of irrigation canals and the distribution of water are fundamental to the productivity of Balinese agriculture. The waters are viewed as coming symbolically from Besakih, the mother temple high on Mount Agung. There is a hierarchy of water temples and small shrines are located in the *sawah* at the junctions of irrigation canals (Lansing 1991). The *subaks*, or irrigation authorities, not only determine the distribution of water and planting schedules; they are also responsible for ceremonies associated with the agricultural round.

Many of the villages in Bali (and there are more than 600 of them) are located on the ridges above the *sawah*. They tend to have a linear form and all have three major temples: the temple of the origin, the *pura desa* or village temple, and the temple of the dead. The former is usually located on the *kaja* side of the village, nearest to Mount Agung, whereas the latter is usually located on the *kelod* or seaward side along with the cemetery. The *pura desa* is usually at the center of the village along with the market, and often a large banyan tree. The temples also have three areas: *jero* (inside, the most sacred), *tengah* (middle) and *jaba* (outside, the least sacred). The entrances to the village are marked by stone pillars or "split gates" as they are sometimes called.

The Balinese usually live as extended families within a walled family compound. Entrance to the compound is gained through an often ornate opening which is backed by a wall which must be circumvented and is designed to confuse and discourage the entry of evil spirits. The open space of the compound is dotted with raised structures or *bales*, each of which is designed to fulfill specific functions and is placed in a particular location in the compound in accordance with Balinese cosmology, with the family temple in the holiest place in the corner nearest to Mount Agung. The *bales* or pavilions are constructed with a roof, uprights and base analogous to the threefold division of head, body and legs. They usually have a thatched roof made of grass (*alang-alang*) and have a crown on their peaks.

The ideas of Balinese cosmology which have been presented above only touch on selected aspects of a complex system of beliefs and their architectural implications. However, sufficient has been written to indicate that the landscape is infused with meaning at a variety of scales according to the application of *Tri Angga*, which informs the Balinese view of the entire island of Bali as well as specific aspects of it, such as the planning of towns and villages, and architectural design. In passing, it may be noted that the notion of *tri hita karana* (the three sources of goodness), encompassing relationships between humans, humans and their gods, and between humans and their environment, is another Balinese trilogy which offers a sound basis for balanced development, with attributes congruent with aspects of sustainable development (Martopo and Mitchell 1995).

The landscape of Bali changes with the seasons as the rice is planted, ripens and is harvested, and with the time of day or night as the sun and moon reflect off the irrigation waters. It is made all the more vivid by the numerous ornate structures in both stone and wood, ceremonies, processions, gamelan music, dance, and displays of craft work and textile products; and one does not usually drive for long in Bali without coming across a colorful procession or other form of celebration.

While many tourists visit temples, drive through an alien landscape, marvel at the agricultural terraces, hear the tinkle of a gamelan, catch snippets of a dance between courses in a restaurant and return home with mementos of these on photographs or video, it is unlikely that most understand more than superficial aspects of what they have witnessed and recorded. As such, they are consumers of a tourism product which is fashioned from a melange of experiences, ranging from the orchestrated event to the serendipitous happening whose touristic meanings and significance may differ substantially from that ascribed to the same features by their Balinese hosts.

Tourism

The official policy of Bali is *pariwisata budaya* or cultural tourism. However, it is evident that visitors to Bali are interested in a mix of natural and cultural attractions. For many, sea—sand—sun attractions predominate, and the rich and vibrant Balinese culture with its associated landscape elements provides little more than an exotic backdrop. For others, the manifestations of culture constitute a more important reason for visiting Bali.

Balinese culture has not been fixed and has evolved in response to outside influences (Vickers 1989). In fact the Balinese pride themselves in their ability to adapt to external pressures and to select those which harmonize with or will enrich Balinese culture, while eschewing others. The tenet *desa, kala, patra* (loosely translated as space, time and condition) suggests that what is appropriate varies with the circumstances. This is explained to the outsider through the analogy with a kite which can be buffeted in all directions by the wind but returns to its original position provided that the string is attached firmly to the ground.

Although small numbers of tourists arrived in Bali early in this century with the ships of the Dutch colonial power, Western visitors in the 1930s had a substantial influence on the image of Bali, and consequently on the form which tourism would take, as well as on the superficial character of some expressions of Balinese culture. Western artists, musicians and devotees of dance, along with noted anthropologists such as Margaret Mead, were attracted by Balinese culture, spread the word concerning its vitality, and in some cases, as in the encouragement of painting, introduced modified forms of expression (Hitchcock and Norris 1995).

The images of paradise which were conveyed were reinforced by the movie industry. Thus the value of Balinese culture was validated by outsiders and it became, among other things, a tourist attraction. This can be seen today in its most developed touristic form in Ubud in south-central Bali, which has grown from a small village to a substantial resort offering many expressions of Balinese culture to tourists.

However, tourism is most highly concentrated elsewhere in the south of Bali in the coastal areas of Badung regency. In the 1960s this area attracted young visitors, especially from Australia, in search of surf, sand and alternative lifestyles (Mabbett 1987). The Bali Tourism Study (BTS), which was undertaken by French consultants with international money and completed in 1971 (Wall and Dibnah 1992), confirmed the southern coast as the main location for tourism and encouraged the further development of the emerging resorts at Kuta and Sanur, and the construction of a new five-star resort complex at Nusa Dua. As well as being the primary areas with white sand beaches, it was felt that concentration of tourists in a limited area in the south of the island would help to minimize the negative impacts of tourism on Balinese culture.

For both good and ill, the Bali Tourism Study largely determined the spatial distribution of tourism in Bali, and has resulted in the transformation of the south of the island, with the construction of hotels and souvenir shops, and the concentration of tourists and associated traffic congestion and pollution of water, air and littoral. In a sense, the south has been sacrificed to maintain the character of the rest of the island. Even so, it was believed by the authors of the BTS that Balinese culture would change irrevocably within fifteen years.

However, with hindsight, this can be seen to have been a simplistic interpretation, for the temples still exist, offerings are made three times each day in front of businesses and in family compounds, and the ceremonies still take place and are often more splendid than in the past as increased prosperity has been plowed back into these communal celebrations. Nevertheless, the landscape has been transformed in Badung regency by the proliferation of hotels, restaurants, souvenir stores and art markets.

At the same time, residents of other areas in the island view the developments in the south with envy and would like to participate in some of the economic activity generated by the growth of tourism (Wall 1996). A recent spatial arrangement plan has responded to these concerns, designating additional areas for tourism development (Wall and Dibnah 1992). These areas are mostly in coastal locations, but together they constitute about one-fifth of the area of Bali. If this plan is enforced, it may lead to a reduction in regional disparities, but at the expense of the more pervasive influence of tourism.

While the construction of tourism facilities is at present highly concentrated, the influence of tourism is less constrained, for tourists are not confined to the resorts but take tours to the interior to experience the manifestations of Balinese culture. This was anticipated by the Bali Tourism Study. Excursion routes were designated to facilitate the movement of tourists from the resorts to the interior during the day and their return at night. In this way they could experience the landscape and culture as well as specific attractions to which the routes gave access but, as has been indicated, it was believed that negative impacts in the interior would be kept to a minimum if visitors spent their nights away from the villages and in the resorts.

This strategy has had considerable implications for the landscape of Bali. The designated attractions, accessed by the excursion routes, have become locations of "art markets" and rough parking lots which are often out of scale with the sites of attraction (more frequently called tourism objects in Bali, for many of them are temples rather than artificial attractions). Furthermore, in the absence of effective building controls, linear development has occurred along the excursion routes. As a result, the views of the *sawah* which they formerly afforded are now less visible, and the viewer interested in landscape appreciation is better served by taking a parallel route.

Indirect effects of tourism are not confined to the resorts. The market in the resorts for fresh fruit and vegetables and other foodstuffs has increased the prosperity of farmers, along with a reduction of subsistence agriculture and an increase in the production of commercial crops, particularly fresh fruits and vegetables. Also, mangrove areas have been converted to shrimp and fish ponds to meet the demand from tourists for fresh seafood, and this, along with the airport extension and port development in Benoa Bay, has made considerable inroads in the last remaining substantial mangrove area.

Fortunately, a proposal to build a water-skiing facility in a "reclaimed" mangrove area has been abandoned and there are attempts to return this and some of the shrimp ponds to their former status. However, there is little experience with the regeneration of mangroves and it is yet to be seen if the initiative is an ecological success or merely a significant gesture.

Wood carving and the production of other souvenirs takes place in many villages throughout the island, though their residents may seldom see tourists, the products being taken to the resorts for sale. Thus some Balinese may benefit from tourism despite little contact with tourists, though it is likely in such situations that the middlemen profit as much, if not more, than the producers. Somewhat similarly, dancers and musicians travel from the villages to the resorts to entertain tourists and earn small sums of money for their efforts.

The spatial layout of most interior villages as described above, remains intact. However, the character of specific villages, such as Ubud, has been changed markedly by the construction of souvenir outlets, restaurants and *losmen* (bed and breakfast establishments), and the signs advertising such services are now ubiquitously displayed in many villages. While Balinese architectural styles have been adopted in tourism developments with considerable success, particularly with respect to the use of carving and stonework, the traditional Balinese village did not possess a hotel or *losmen*, or any of the other trappings of tourism, nor did it usually have two-storey buildings (such constructions are a relatively new architectural feature in the interior of the island).

Indeed, though the Balinese compound, with its open spaces and *bale*, can be readily converted for supplementary tourism uses, it is only with the possible loss of religious significance (Wall and Long 1996). Consequently, where the household temple is raised to the second floor to retain its *kaja* position, special efforts have to be made to ensure that it maintains its link to the earth (Sulistyawati 1989). Furthermore, in some of the older resorts, *alang-alang* is being replaced with tile as a roofing material.

Thus tourism has brought new forms to the landscape, ranging in scale from huge new resort complexes such as Nusa Dua, to the proliferation of smaller tourism establishments, particularly in Kuta but also in many interior villages, and including the numerous small signs advertising specific outlets. Indeed, one temple even has a carving of a motorbike—now a common means of transportation in Bali—inscribed on its walls!

Perhaps of more concern, though, than an isolated temple carving is the construction of large, new up-market accommodation establishments on valley sides in the interior of the island. The sites have been chosen because of the magnificent views which they provide across the *sawah*. However, there is a danger that they will compete for water, which is needed to irrigate the rice, which in turn is an important component of the vista and a rationale for their location. Furthermore, should waste disposal prove to be inadequate, further stresses will be placed on the *sawah* below and downstream. In coastal areas such as Sanur, inadequate waste disposal, coupled with mining for building materials, has already destroyed coral reefs. As a consequence, erosion along the shoreline is now severe, resulting in the construction of unsightly offshore barriers and rip-rap (a foundation or sustaining wall of stones intended to prevent sand attrition).

The beach as a contested landscape

It has been argued that the Balinese have traditionally turned their back on the ocean and oriented their activities towards the interior. Nevertheless, the beach has customarily been a place for the conduct of certain ceremonies as well as a place to cleanse oneself in the shallow water. In some locations it is a convenient site on which to land and store boats or, as on the island of Nusa Lembongan, a place to dry seaweed, the product of seaweed farming (Long and Wall 1996). For Balinese children it may be a good place to fly a kite in the onshore or offshore breeze, while tourists, who now dominate the white sand beaches, find it an enjoyable place to relax, to sunbathe or to take a leisurely walk. Other more active participants utilize it as a base from which to parasail, jet-ski or bungee jump, while investors, both Balinese as well as international, find that it presents a business opportunity and a chance for a speculative venture.

Indeed, the beach is an active workplace for many in the informal sector who roam the beaches (excepting Nusa Dua where the practice is prohibited) to sell postcards, jewelry, watches, carvings or clothing, or to offer services such as hair-braiding and massage. For the most part, they seek business in proximity to large hotels which are often under the control of off-island investors. Many are themselves immigrants, particularly from Java, although the unsuspecting international visitor is usually unable to tell the difference between Balinese and Javanese salespersons and, lacking a reason for thinking otherwise, assumes that they are the former (Cukier-Snow and Wall 1994).

While the linear developments of large hotels along the coast have greatly reduced beach accessibility for many customary users, the break of slope on the beach is also used to physically separate the tourists from the traders, and many hotels have guards who will intervene to protect the tourists from over-zealous vendors who cross the line in their search for a sale. Thus the beach provides a good example of the way in which landscape elements have been re-evaluated and are viewed and used differently by different people reflecting different purposes as well as differential access to power.

Planning implications

Balinese landscape and culture is a major Indonesian tourism asset. At the national scale, Bali is viewed as a major attraction which can generate foreign exchange for the benefit of the national economy. In particular, it is the hub in a "hub and spokes" strategy by which visitors are encouraged to come to Bali and then move on to explore other areas in Indonesia (Wall 1997a).

Meanwhile, some Balinese would like to see more balanced development through a mix of tourism, agriculture and small industry, and prefer a greater emphasis on encouraging longer lengths of stay on the island rather than the development of attractions and destinations elsewhere.

Balinese dance and dancers, the ornate woodcarvings and the cascading rice fields have become national symbols, even though the Balinese constitute a minority in Indonesia, much as Inuit carvings have become a symbol of Canada. However, construction is underway of a giant garuda (a mythical bird) and associated tourism attractions on the Bukit Peninsula on the southern tip of the island, ostensibly because Bali is in need of a symbol. Constructed on the scale of the Statue of Liberty, it is being built in a limestone area which is poorly served by infrastructure and thus likely to provide challenges for efficient and effective water supply and waste disposal.

At the same time, large accommodation developments are taking place adjacent to Tanah Lot, one of the most important temple sites in the south of the island. This development has been the source of open dissent, an unusual situation in a society which thrives on consensus and in which criticism is usually muted (Cohen 1994). Clearly the official policy of *pariwisata budaya* (cultural tourism) is not always interpreted in a consistent manner, and it is not difficult to find dissonance between the rhetoric and the reality. Nevertheless, aspects of Balinese culture, including *Tri Angga*, have the potential to inform tourism planning if they are allowed to do so.

I have suggested elsewhere that attraction type (cultural, natural and recreational), location (the interior and the coast), the spatial characteristics of the resource base (nodal, linear and areal) and status of development (whether highly developed or pristine) could prove useful in specifying appropriate forms of tourism in particular geographic places (Wall 1993, 1997b). Thus recreational developments might be concentrated in selected coastal locations, as the major ones currently are, and discouraged in the culturally important interior. Furthermore, such strategies could be implemented through specification of the numbers and types of accommodations best suited to a particular site, reflecting its environmental and cultural sensitivity.

Conclusion

Although tourism is concentrated in the south of Bali in Badung regency, few parts of the island are devoid of its influence. Even where tourists are few, goods are produced for the tourist market and people migrate to the resorts in search of employment. Using dance as an example, the French scholar Picard (1992) has called Bali "un musée vivant" (a living museum) and has used the word "touristification" to describe a situation in which tourism is so pervasive that it is no longer possible to separate that which is indigenous from that which is attributable to tourism.

He points out that dances formerly performed for the gods are now performed in shortened and simplified forms for the tourists. Conversely, dances created originally for the tourists are now performed in the temples. Offerings to the gods are usually made in both cases, although I suspect that most dancers are aware of when they are dancing primarily for the gods and when the audience is tourists.

The paving of roads, increased motorcycle and automobile ownership, rural electrification, a successful family planning program, and the agricultural "green revolution" are other aspects of modernization experienced by Bali recently. Not all of the changes in the Balinese landscape and culture should be ascribed to tourism, however, for it is impossible to separate the role of tourism from other forces of change. Nor is this necessarily desirable, for they occur together and not as discrete phenomena.

Certainly, Bali is changing, as is the case with any vibrant culture. However, the changes wrought by tourism may be evaluated differently when viewed from a national or a local perspective. Furthermore, the same policy or development is likely to have different implications for different segments of the population, such as women and men (Cukier et al. 1996). As an outsider, it is extremely important to distinguish between those changes which may be welcomed and those deemed undesirable and which will therefore be resisted, for many Balinese desire the material benefits of a higher standard of living and increased life opportunities for their children.

Instead, the landscape of Bali may be best viewed as a palimpsest of layered, highly individualized meanings. Tourism, in its varied forms, adds new elements and users to the landscape, and gives added value to the physical landscape and culture of Bali. As such, the physiographic and human landscapes are continually (re)created, modified and shared, at a price, by, among others, the Balinese, their gods, the tourism industry and the tourists. The resulting destination is neither an expression of Balinese culture nor an artifact of tourism alone. Rather, it is an amalgam of a multiplicity of influences which have combined to form a changing melange that is now distinctively Balinese. As such, this process of "place creation" merits further attention.

References

Budihardjo, E. (1986) *Architectural Conservation in Bali*, Yogyakarta, Indonesia: Gadjah Mada University Press.

Cohen, M. (1994) "God and mammon: luxury resort triggers outcry over Bali's future," *Far Eastern Economic Review*, 26 May: 28–33.

Cukier J., Norris, J. and Wall, G. (1996) "The involvement of women in the tourism industry of Bali, Indonesia," *The Journal of Development Studies*, 33, 2: 248–70.

Cukier-Snow, J. and Wall, G. (1994) "Informal tourism employment in Bali, Indonesia," *Tourism Management*, 15, 6: 464–7.

Hitchcock, M. and Norris, L. (1995) *Bali, The Imaginary Museum: The Photographs of Walter Spies and Beryl de Zoete*, Oxford: Oxford University Press.

Hull, R. B. and Revell, G. R. B. (1989) "Cross-cultural comparison of landscape beauty evaluations: a case study in Bali," *Journal of Environmental Psychology*, 9: 177–91.

Knudson, D. M., Cable, T. T. and Beck, L. (1995) *Interpretation of Cultural and Natural Resources*, State College PA: Venture Publishing.

Lansing, J. S. (1991) *Priests and Programmers: Technologies of Power in the Engineered Landscape of Bali*, Princeton NJ: Princeton University Press.

Long, V. and Wall, G. (1996) "Successful tourism in Nusa Lembongan, Indonesia?" *Tourism Management*, 17, 1: 43–50.

Mabbett, H. (1987) *In Praise of Kuta*, Wellington, New Zealand: January Books.

Martopo, S. and Mitchell, B. (eds) (1995) *Bali: Balancing Environment, Economy and Culture*, Waterloo, Canada: Department of Geography, University of Waterloo.

Micheaux, B. (1994) "Land movements and animal distributions in east Wallacea (eastern Indonesia, Papua New Guinea and Melanesia)," *Palaeogeography, Palaeoclimatology, Palaeoecology*, 112: 323–43.

Nash, D. (1989) "Tourism as a form of imperialism," in V. Smith (ed.) *Hosts and Guests: The Anthropology of Tourism*, 2nd ed., Philadelphia PA: University of Pennsylvania Press, 37–52.

Picard, M. (1992) *Bali: Tourisme Culturel et culture touristique*, Paris: L'Harmattan.

Sulistyawati (1989) "The Balinese home: factors that influence change in its architecture," unpublished MA thesis, University of Indonesia.

Vickers, A. (1989) *Bali: A Paradise Created*, Berkeley CA and Singapore: Periplus.

Wall, G. (1993) "Towards a tourism typology," in J. G. Nelson, R. W. Butler and G. Wall (eds) *Tourism and Sustainable Development: Monitoring, Planning, Managing*, Waterloo, Canada: Department of Geography, University of Waterloo.

——(1996) "Perspectives on tourism in selected Balinese villages," *Annals of Tourism Research*, 23, 1: 123–37.

——(1997a) "Indonesia: the impact of regionalization," in F. M. Go and C. L. Jenkins (eds) *Tourism and Economic Development in Asia and Australasia*, London: Cassell, 138–49.

——(1997b) "Tourist attractions: points, lines and areas," *Annals of Tourism Research*, 24, 1: 240–3.

Wall, G. and Dibnah, S. (1992) "The changing status of tourism in Bali, Indonesia," in C. P. Cooper and A. Lockwood (eds) *Progress in Tourism, Recreation and Hospitality Management*, vol. 4, London: Belhaven Press, 120–30.

Wall, G. and Long, V. (1996) "Balinese homestays: an indigenous response to tourism opportunities," in R. Butler and T. Hinch (eds) *Tourism and Indigenous People*, London: International Thomson Business Press, 27–48.

World Tourism Organization (1997) *Tourism Highlights 1996*, Madrid: WTO.

4

TOURISM EMPLOYMENT AND SHIFTS IN THE DETERMINATION OF SOCIAL STATUS IN BALI

The case of the "guide"

Judith Cukier

Tourism employment has the potential to effect cultural change within a community, as well as cause a shift in social status for those employed. It can modify social organizations and cause changes in cultural values that, in turn, may have implications for the determination of social status (Cukier-Snow and Wall 1993). Guiding is a particularly relevant example of tourism employment activity which demonstrates how such employment may result in the emergence of high-status occupations acceptable to the local society.

Guiding is a service provided to tourists by certain members of the local community. In addition to providing basic transportation services, guides require a knowledge of local customs, culturally important sites and societal organizations. Thus guides are often seen as "cultural emissaries," providing a link between the tourists and the local community. Guiding is also an occupation that is both demanding and rewarding in all senses of the words. As a result, societal recognition of these job characteristics can result in guides being accorded a relatively high social status.

This chapter begins with a review of the literature on tourism employment and the effect tourism employment has on social status. After introducing Bali, Indonesia, as a tourism employment case-study, the chapter then goes on to situate guiding within the broader context of tourism employment, and presents the results of interviews conducted with guides in two coastal resort villages in Bali: Sanur and Kuta. It concludes with an examination of the personal, societal and policy implications of tourism employment in Bali.

The status of tourism employment

In developed countries, employment in the tourism sector is normally considered to be low-status. This is partly due to the low remuneration of the sector when compared to others, and to the lack of emphasis that many developed-country cultures place on the importance and value of the "service" sector. Turner (1973), commenting on developed countries, referred to tourism as the most subservient industry, and argued along with Urry (1990) that tourism work was servile in nature. This view has persisted for decades. For example, Hudson and Townsend (1992) also argued that tourism employment in Britain was characterized by low-quality work and was accorded an associated low status.

Some researchers have transferred the status of tourism employment in developed countries to developing countries, often without any empirical basis (Turner 1973; Young 1973; Diamond 1977; Duffield 1982; Vaughan and Long 1982; Richter 1986; Bagguley 1987; Rajotte 1987; Pigram 1990; Baum 1993). However, the servile nature and low remuneration attributed to tourism occupations in the developed-country literature may not be representative of developing countries. Nonetheless, tourism employment in such areas is thought by some to be of low status and therefore degrading and demeaning. Elkan (1975: 124) adopted this view with reference to tourism employment in Africa. He described the Masai as "one time proud [people who] now demean themselves to selling trinkets and begging for sweets." However, his conclusion reflected his own interpretation of the situation rather than that of the Masai and their view of the status of their newfound positions.

In contrast to the status of tourism employment in developed countries, opportunities for work in the tourism sector in developing countries are often accorded a relatively high status by the local population. Reasons include the relatively high salary of tourism occupations (even those within the informal sector), the relative physical ease of many tourism occupations when compared to traditional primary sector occupations (e.g. agricultural labor), and the security of some occupations and the flexibility of others. In addition, perspectives on the servile nature often associated with tourism occupations may be strongly culturally based. For example, in many developing countries service toward others—and the service industry in general —is allocated a high status and importance in society.

In Southeast Asia in particular, local cultures place a high value on "service" toward others, a situation reflected in the relatively high status accorded to tourism occupations. The potential for change in social status and associated shifts in attitudes was speculated upon in a UNESCO (1976) report on the socio-cultural impacts of tourism. The report identified two traditional employment sectors, fishing and agriculture, whose employees would undergo considerable social change with a shift to tourism employment, primarily due to an increase in social status.

Subsequent to the UNESCO report, others have examined the status of tourism occupations in developing countries. Pongsapich (1982) found that in Thailand, young people who were fluent in English or other foreign languages often became tour guides, a position that was considered to be a good job in terms of both pay and status. She went on to state that "guiding," an informal job, ranked higher in status than many other formal tourism occupations such as bussing tables, washing dishes or working as a janitor.

Pizam *et al.* (1994) compared the perceptions of tourism among tourism employees in both Florida and Fiji. The authors noted that, overall, Fijian residents accorded a higher status to employment in the tourism industry than Floridians, a finding the authors attributed to the greater availability of well-paid options in Florida compared to the more limited and less lucrative alternative employment opportunities in Fiji.

Occasionally, the opinion of researchers regarding the status of tourism employment in developing countries is ambiguous. Lovel and Feuerstein (1992) stated that most people working in tourism in New Caledonia, Fiji and Hawaii were given low-status positions such as waiters, clerks, maintenance staff and maids. However, in the same article the authors commented that jobs in tourism often paid better than any alternatives, resulting in labor being attracted from other sectors into the tourism sector. Since high status and high salaries are often related, a high-salary occupation is often accorded a high status. Lovel and Feuerstein provided the example of nurses in the Pacific region who left their jobs to work in the tourism sector. This complexity regarding the true status of tourism employment suggests that more research is needed.

Tourism and status: a shift from traditional ways?

Tourism employment, like any relatively new employment sector, has introduced a new means for determining social status. In an article examining gender and tourism issues in South Java, Wilkinson and Pratiwi (1995) argued that social stratification in Indonesia is increasingly taking on an economic basis as opposed to the traditional class stratification system based on position within the state apparatus (Gordon 1978) and culture. In their case-study of the Pangandaran region, people who were considered by their peers to be "upper class" were employed in the formal tourism sector or owned tourist accommodations and restaurants, while "lower class" individuals were employed in the informal tourism sector.

Unlike most other employment sectors, both formal and informal tourism occupations may lead to gains in social-status contributions to individual and community prosperity and development. A number of researchers (Pongsapich 1982; Ebery and Forbes 1985; Lovel and Feuerstein 1992; Pizam *et al.* 1994) have argued that informal sector tourism workers acquired a higher status because of the skill, initiative and relatively high remuneration associated with their occupations. The high status associated with tourism employment results in self-fulfilment and self-actualization of its employees.

In addition, informal sector employment may be a crucial training ground that leads to cross-overs to the formal sector, thus increasing personal development through increased status, skills and education. Within the informal sector there is a growing differential in socio-economic status among workers. Ebery and Forbes (1985), for example, argued that some informal sector street vendors were acquiring a higher status than some formal sector employees because their enterprise required skill, initiative and capital investment. The acquisition of skills, whether through formal or informal means, results in a population with skills in demand at the higher levels of tourism employment. Consequently, over the long term, there should be reduced reliance on the use of "outsiders" to fill higher-level, managerial positions. Thus tourism employment is beneficial to both the formal and informal economic sectors, and contributes economically and developmentally to both individuals and the community.

Tourism in Bali, Indonesia

The island of Bali is one of the most touristed of Indonesia's provinces, with more than one million tourists (approximately a quarter of all tourists to Indonesia) visiting the island in 1995 (Cukier et al. 1996). Although travelers visited the island prior to the twentieth century, Bali did not become a prominent global destination until the late 1960s, with the construction of an international airport and the island's first five-star hotel. The population of the island in 1990 was 2.8 million, with an annual growth rate of 1.2 percent and a density in 1991 of 500 persons per square kilometer. Nearly a quarter of the residents live in the southern coastal areas (UNDP 1992) and, unlike predominantly Moslem Indonesia, the main religion on Bali is Hindu.

One significant result of the rapid development of tourism in Bali has been dramatic economic growth, with the island experiencing one of the highest average income levels in all of Indonesia (Cukier-Snow and Wall 1993). Approximately 20 percent of the Gross Provincial Product in 1990 was derived either directly or indirectly from tourism, and employment in tourism has increased dramatically in both the formal and informal sectors (Picard 1992).

Though the exact number is as yet undetermined, a substantial number of people in Bali are now employed in tourist-related enterprises and, since 1970 and the initiation of mass tourism, many people from all over Bali and other Indonesian provinces have left low-wage traditional sectors such as farming and fishing, and migrated to the main resort areas of Kuta and Sanur in order to seek employment in the tourism industry. One highly visible result of this growth has been a rapidly changing landscape as more hotels, restaurants and souvenir shops are built.

As if in support of this growth, the Balinese government has adopted the position that tourism in Bali must be *pariwisata budaya*, or "cultural tourism" (Picard 1990), meaning that tourism must be respectful of the cultural values and artistic traditions that initially attract visitors to the island. This policy has been successful insofar as it has raised awareness among the Balinese to the issue of "cultural identity." Furthermore, the tourism boom has had some positive cultural effects.

Money spent by visitors has, in some cases, enhanced an interest among Balinese in traditions such as craft production and traditional dance that may have otherwise lost some significance (Picard 1992). Even more important may be the positive influence of the admiration of and interest in Bali and Balinese culture expressed by visitors. This has almost certainly led to the strengthening of the Balinese sense of cultural identity and pride.

Tourism employment and the determination of status

Many Balinese scholars as well as individuals involved in the tourism sector in Bali believe that tourism employment has replaced the traditional caste system as an indicator of status (Manuaba 1991; Geriya 1993; Mandra 1993; Sudiana 1993; Sukabrata 1993). Everyone in Bali is born into a caste, with the majority of the population belonging to the common *sudra* caste. Members of the minority belong to one of the three remaining gentry castes: *brahman* (priests), *satria* (nobles), or *weisa* (merchants) (Geertz and Geertz 1975). According to Geertz (1973), the caste system dictates religious status and does not affect moral, economic or political status.

Wayan Geriya (1993), a Balinese cultural scholar, has argued that tourism in Bali has caused a shift in status determination. Where previously a person's status was determined by birth caste, it is now more frequently based upon achievement and type of employment, with the tourism sector providing ample opportunity for both high achievement and prestigious employment. Vickers (1989) supported this view with his argument that poverty in Bali had become associated with a lack of access to tourism income.

The determination of social status in Bali as it relates to tourism is not just a matter of a high salary, but also of the education needed to land the job, the association of the individual with a "name brand" of tourism product, and the skills needed or acquired in order to perform the job. By way of example, hotel work in Bali is accorded a high status in part due to the high salaries paid: one five-star hotel in Kuta pays a minimum monthly salary of 325,500 rupiahs,[1] rising to 1,000,000 rupiahs for senior employees (Sukabrata 1993).

According to the Director of the Hotel and Tourism Training Institute of Bali (BPLP),[2] I. Nyoman Bagiarta, their tourism training degree program attracts Balinese of a higher caste who are not interested in traditional diploma-granting programs. For these people, he argues, money is not the only important factor; recognition of a degree and status are also important.

For example, work in hotels is accorded a high status, especially if the hotel is large or part of a recognized international chain (Bagiarta 1991). A reflection of the remnants of colonial influence, the ability to speak a foreign language is also accorded a high status (Sudiana 1993; Sukabrata 1993).

Guiding as a tourism employer

The role of the guide in Bali is quite varied and can range from simply driving tourists to various sites of interest, to acting as a commentator and interpreter of the destination's culture. In this sense, guides can be seen as "cultural emissaries." This being the case, Akroun (1992) commented that guides should be trained and compensated as high-level professionals, since they serve as mediators of the cultural identity of their country when introducing tourists to their cultures and traditions.

Gertler (1992) went even further, arguing that "tourism [which is] based on cultural heritage demands a superbly high standard of tourist guiding." He commented that

> since the guide's motivation is to inspire the tourist's interest by delving more deeply into the culture itself . . . tourist guiding becomes a genuine learning process. It becomes a potent revitalizing force in an upwardly spiraling feedback process that progressively enhances the positive impact of tourism, culturally, as well as economically.
>
> (Gertler 1992: 15)

Thus guiding should not be considered a menial activity, but rather a desirable and culturally important occupation which can positively influence both the individual employed within it and the larger community. Guiding may be considered as both a formal and informal sector activity, with variation in the degree of formality. Bromley and Gerry (1979), in attempting to classify types of employment, depicted employment categories along a continuum stretching from the least autonomous, "stable wage work," to the most autonomous, "true self-employment." For example, a guide owning a vehicle would fit within the "true self-employment" category, defined as someone who works independently, has free choice of suppliers and outlets, and is the sole owner of the means of production.

A guide who is "registered" as a licensed guide, yet uses a vehicle owned by a transportation company or hotel, would be categorized by Bromley and Gerry (1979) as "disguised wage work." Under this heading, a firm or group of firms appropriate part of a worker's product without that person legally being an employee. Rather, they are classified as a subcontractor or commissioned salesperson. These guides would be considered part of the formal employment sector.

Finally, an "unregistered" guide operating someone else's vehicle and keeping the net profit, would fit into Bromley and Gerry's "dependent work" category. Here the worker must rely upon one or more larger enterprises for credit and/or the rental of premises or equipment. Appropriation of part of the product is not as direct as with the disguised wage worker, but takes place through payment of rent or repayment of credit. Guides who are part of this category fit somewhere along the continuum between the formal and informal sectors.

However, it is interesting to note that the degree of employment formality does not necessarily reflect wage earnings. For example, a guide who is truly self-employed and is part of the informal sector may have a higher income than the formal-sector "disguised wage worker."

Guiding in Bali

In Bali, "guides" are those individuals with access to a vehicle, and who take tourists to a variety of tourism sites around the island while possibly, but not necessarily, providing some level of commentary. Some guides belong to an official organization (or licensing body), while others represent a tour company, work as freelance drivers or are hired by tourists "off the street." These freelance guides attract tourists by initiating contact with them and often suggest sites in Bali which may be of interest to tourists.

As tourism to Bali has increased, so has the demand for qualified guides. In 1974, the Bali Guide Association estimated that demand during the peak tourist season exceeded the 120 guides then available, but were unable to project how many more were needed. By 1985 the number of official guides registered was 487, and in 1990 the number had grown to approximately 1,350[3] (Bali Government Tourism Office 1990). By 1993 there were 2,398 registered guides employed in Bali, of whom 190 were women (BGTO 1993).

The under-representation of female guides in Bali was explained by the 1992 UNDP study as largely a consequence of socialization factors. Traditionally, Balinese women are chaperoned when in the company of men and guiding contradicts this custom. In addition, two related reasons make this occupation an uncommon choice for women: females are not generally encouraged to drive, nor to venture far from home without friends or family (while women do drive in Bali when alone or with another woman, they will usually abstain when in the company of a man).

Ngurah Wijaya, of the Government Guiding Regulation Department, estimated that 90 percent of the registered guides in Bali in 1993 were Balinese (Wijaya 1993). Despite the fact that guides act as "culture brokers" for tourists, registered guides need not be Balinese, as long as they meet the requirements to work within the guiding sector, which, among other things, requires knowledge of Balinese culture.

Guiding has evoked a negative image in Bali. A 1993 *Bali Post*[4] article estimated that a total of 5,000 guides, including those who were unregistered (constituting the unofficial or informal sector) operated in Bali. The presence of "informal sector" guides was considered by the Government Guiding Regulation Department to be creating a negative image for tourism in Bali. Much of the perceived problem is due to high commissions (usually between 40 and 60 percent) paid to guides by art shop owners.

It has been difficult to control this system of "kickbacks" and to determine whether it is carried out by registered or unregistered guides (Picard 1992). A typical sales breakdown would be a 10 percent commission to the driver, 20 percent to the guide, 20 percent to the transportation company, and the remainder (50 percent) to the art shop. This system is too expensive for smaller art shops which cannot pay these "kickbacks" and instead rely increasingly on mobile vendors to market their products (Manuaba 1993). As a result tourists pay higher prices for goods as well as services (such as guiding) in Bali.

Vickers (1989) has argued that guides are now one of the most prosperous "middle class" groups in Bali and guiding is now a goal to which many Balinese aspire. Whether registered or not, to be a guide is to be employed in one of the more prosperous tourism sectors in Bali, and higher salaries, according to Vickers, can be generated by guides through their interaction with art shop owners and other major figures in the tourism sector. Most important, however, guides earn a comparatively high salary simply due to the fact that most international tourists have the financial means and willingness to pay top dollar for a high-quality guiding service.

Case-study: interviews with guides in Sanur and Kuta

Sanur and Kuta were selected as case-study sites because of their prominence as tourist resorts in Bali, as well as their differing resort characteristics. Sanur is a large village situated on Bali's southeast coast about six kilometers from the capital city of Denpasar. Before tourism arrived in Sanur[5] residents worked predominantly in farming, fishing, animal husbandry and artisanal activities (Picard 1993). Although Sanur residents were able to support themselves economically by these traditional means, the area was relatively poor, largely due to the limited availability of arable land.

Thus the introduction of mass tourism to the area in the late 1960s was seen favorably by local residents (Lindayati and Nelson 1995). A tourist beach market was created, restaurants and art shops were opened by local villagers, and by the late 1980s over half of Sanur's working population were employed in tourism-related activities (Lindayati and Nelson 1995). A similar development pattern occurred in Kuta, where pre-tourism employment consisted primarily of fishing and farming, with very limited craft development (Hussey 1989). The first tourism employees in Kuta were entrepreneurs who set up small-scale, locally owned accommodation, restaurants, clothing and souvenir shops, and bicycle and motorcycle rentals. Today tourism is the most visible economic sector in Kuta, with tourism-related enterprises dominating the landscape.

The following section presents the results of sixty structured interviews (thirty in Sanur and thirty in Kuta) with drivers/guides who take tourists on tours around the island. The interviews were conducted verbally by Balinese research assistants and each took approximately forty-five minutes to complete. Respondents were approached randomly, and thus no effort was made to seek out an equal number of male and female respondents.

General demographics of guides

Eighty-seven percent of the guides interviewed were Balinese, with the remainder originating from the neighboring island of Java. Forty percent of the guides were less than thirty years of age, though the average age of the groups was 33.7 years. Sixty-five percent of those sampled were married and approximately a third of them (31 percent) reported that their spouse worked as a "housewife," while an additional 28 percent reported spouses employed as vendors selling either products or services to tourists.

The majority (55 percent) of guides earned more than Rp400,000 per month, with approximately a third (32 percent) earning over Rp500,000 per month. Most borrowed funds to purchase their vehicle and reported long working hours (more than eight hours a day), indicating that they chose their work for monetary reasons. Universally, the guides indicated that they spoke at least some English, and 43 percent indicated that they could speak at least one other foreign language in addition. All stated that they had learned foreign languages "on the job" through communication with tourists.

Although other studies have demonstrated that many tourism employees in Bali are in fact migrants from other provinces in Indonesia (Cukier-Snow and Wall 1994), only 13 percent of the guides interviewed were non-Balinese. In part, this is probably because working as a guide is an occupation in which local knowledge of traditions and Balinese culture is beneficial, as is a detailed knowledge of Bali's tourist sites. New migrants to the island do not usually have this knowledge and thus initially choose alternative employment such as beach or street vending.

Job satisfaction and status

Several indicators of job satisfaction were evaluated through the interviews, including respondents' long-term career aspirations, desirable employment alternatives, and reasons for initially choosing their current tourism job. The majority (83 percent) indicated that they were content with their current position and desired to remain in the same capacity over the long term. Asked to suggest three reasons why they liked their occupation, most (73 percent) cited the satisfaction derived from their work and an associated improvement in the general quality of life. Slightly less than a third (27 percent) stated that they liked their job because they felt they had no other option available to them. For this group, guiding was deemed preferable to unemployment. Only one guide attributed his job enjoyment to the high status conferred on it.

In addition to the guide interviews, tourism workers who were not employed as guides (e.g. hotel employees, vendors) were asked to identify up to three high-status tourism occupations. The majority of interviewees, regardless of birthplace, place of work or gender, believed that either hotel employees or guides were the tourism occupations accorded the highest status within the tourism sector. Several interviewees indicated that tourism occupations in which the employee was required to wear a uniform were high-status occupations. Since guides often wear uniforms, they are seen by others as high-status individuals.

In general, perceived status was not strictly linked to income levels, depending also on foreign language ability, the "honor" of wearing a uniform, the size of the tourism employment venture, and the type of tourism employment. The following "life histories" of two working guides are intended to provide the reader with a greater understanding of some of the characteristics of guiding in Bali.

Life-history excerpt: a female guide

Suarti is a twenty-six year old female guide working for a transportation company in Sanur. She was born in Gianyar, Bali, and still lives in, and commutes daily from her home village, about thirty minutes drive from Sanur. Suarti always wanted to be a guide, mainly because she grew up surrounded by tourism, and felt that by becoming a guide she could help tourists gain a better appreciation of Bali. She believes that Balinese culture will not be destroyed by tourism; rather, she sees her work as an ideal way to strengthen the culture by fostering its appreciation among foreign visitors.

After graduating from high school, Suarti took a three-month English language course followed by a similar course in guiding. In June 1992 she received her "official" guiding license from the government and has been employed as a guide ever since. Suarti lives with her husband and his family and attends most of the important village and temple ceremonies. However, her parents-in-law often represent her and her husband at the less important village functions, and Suarti and her husband are often required to pay a fee to the village to compensate for their absence. Suarti can request time off for temple or village ceremonies, but she must give at least two days notice to her employer.

Though the majority of guides in Bali are male (in the company where Suarti works there are forty-two male guides and only three females), women are shown no favoritism and must work at night if required. Suarti does not resent the competition from "unofficial" guides since, having worked as a street vendor prior to becoming a guide, she understands that the need to work may override government requirements for official licensing.

Life-history excerpt: an "informal" guide

Made is a forty-five year old guide and native of Singapadu, Bali. He stated that ever since he was a child, his dream was to become a driver for tourists. Before acquiring his present job, he was an assistant bus driver on the route from Bali to Java. In 1975 he became a taxi driver for the Bali Beach Hotel, and subsequently worked as a driver for the Sanur Hyatt Hotel. Made is employed by a transportation company which leases the vehicle from a three-star hotel in Sanur.

Although he works as an employee, he is considered an "informal" or "unofficial" guide, since he does not have the required government guiding license. His monthly salary from the transportation company is Rp40,000 (US$20), but he also receives a percentage of the transportation company's profit, thus increasing his salary to at least Rp240,000 ($120) per month.

Made has never had any formal foreign language training, although he learns languages by listening to and speaking with other guides and tourists. He generally takes days off to attend temple and village ceremonies, but must give his employer at least two days prior notice. Made is married and has four children, the oldest of which is studying at the Hotel and Tourism Training Institute of Bali (BPLP). His remaining three children are still in school and his wife works as a farmer.

Cultural life

One of the goals of the case-study was to determine if cultural aspects of Balinese life were being significantly affected by the shift away from traditional sectors such as farming and fishing toward tourism employment. Tourism employment was found to be a relatively recent phenomenon, with the parents of most interviewees having worked in traditional occupations such as farming or fishing, or in the case of mothers, as "housewives." Only a small proportion (16 percent) of interviewees' parents were employed within the tourism sector, reflecting within one generation, both the recent development of tourism in Bali and the resulting changes in the occupational structure.

In contrast, over half of the interviewees' spouses worked within the tourism sector, and many indicated that they believed tourism-sector work would be a desirable career choice for their children. Parent-child employment pattern shifts is one indicator of cultural change, and in Bali this pattern is found in the shift from traditional-sector employment to more "modern" tourism employment. It is apparent that occupations in tourism are increasingly being favorably viewed, and that a shift away from traditional sectors will accompany continued tourism development in Bali.

Socio-cultural implications

The introduction of tourism employment to Bali has had economic, social and cultural implications. Overall, the economic implications have been positive, through the raising of individual income levels, the generation of employment in industries linked to tourism (e.g. handicraft and garments), and overall infrastructure improvement. However, it is not clear whether the overall increase in community income contributes to improved community welfare. According to Geriya (1993), in order to guarantee improved community welfare, tourism income should contribute more broadly to the enhancement of education, the maintenance of the environment and the continuation of religious rituals.

The social implications of tourism employment are more difficult to assess than the economic implications. According to some scholars (Bagiarta 1993; Geriya 1993), social networks in Bali have been widened as a result of tourism employment. People who work in agriculture are tied to land within their village and thus associate predominantly with other villagers. In contrast, tourism employees such as guides originate from a number of villages and are exposed to many new people, including foreign tourists and Indonesians from other provinces.

Research undertaken in Bali suggests that people who have moved from their home village in order to take up tourism employment do not return to their village for all temple ceremonies, nor to participate in normal community activities such as *gotong royong* (community cleaning/maintenance). Instead, these tourism employees pay a fine to compensate for their absence or are represented by family members. Some guides reported that they had sold their land in order to pay for their vehicle, a finding which has major implications for the maintenance of Balinese culture, since so much of the religion and culture of Bali is tied to agricultural land through the *subak* (agricultural maintenance group) system (Lindayati and Nelson 1995).

Balinese culture is still very strong, even among the youth. It is not uncommon to see a guide leave work and, wearing a traditional outfit adorned with a leather jacket, hop on his/her motorcycle and head off to a temple or village ceremony. Tourism employment has not radically altered traditional customs in Bali. However, because of the relative infancy of tourism in Bali, only one generation has been raised surrounded by and employed within tourism. When the young people who are finding work in the tourism sector return home they are confronted with the traditional customs maintained by their parents and grandparents.

A key question is whether—and in what manner—these traditions and customs will be carried on and passed down to future generations. Francillon (1990) optimistically asserted that the strength of the people of Bali, their social cohesion and ability to cope with recurrent natural, or other, disruptions have allowed them to resist, and recover from floods, earthquake, and volcanic eruptions many times in the past. Similarly, they could resist, and possibly absorb, tourism.

(Francillon 1990: 271)

Policy implications

The research suggests that official government policies promoting cultural tourism should acknowledge the importance of non-Balinese and informal-sector tourism employees in the formation of tourists' cultural experiences. Thus these individuals should be incorporated into the cultural tourism policy framework. Tourism in Bali is being promoted as "cultural tourism," yet many tourism employees are non-Balinese. For an Indonesian island which promotes cultural tourism, this finding is significant to government and tourism planners.

Guides in Bali act as "culture brokers" for tourists, yet many guides are "unofficial," unregistered, and consequently have received no appropriate training. Thus a more effective policy, targeting both "unofficial" guides and non-Balinese tourism employees, would recognize these tourism employees as "cultural emissaries" and provide them with free or low-cost training courses in Balinese culture, customs and traditions.

In contrast to traditional employment theory, which states that the informal sector of developing countries will "disappear" as a country achieves "development," more modern views assert that this dynamic sector should not be simplistically viewed as one which will vanish with the processes of modernization and urbanization. Within tourism, as in other industries in developing countries, the informal sector is a vigorous and notable element, and thus efforts should be made to better understand informal-sector partici-pation in tourism and its relationship to the formal sector.

One policy specifically geared to the informal tourism sector would be the provision of low-cost or free courses in entrepreneurship, which would assist this large and viable tourism sector in acquiring skills such as marketing, finance, foreign languages and investment knowledge. Such courses as have been initiated for informal-sector tourism employees in Yogyakarta, Java, would be particularly beneficial in Bali, where the informal sector is so promi-nent.

Another policy implication arising from the research is support for tourism education and training institutes. A deliberate policy to provide a high-quality general education as well as specific tourism training will help ensure that the local population will gain access to a variety of higher-level manage-ment positions within the tourism sector. Blanton (1981: 120) argued that the existence of formal tourism training institutes results in the elimination of the view that the tourism-sector employee is a "marginal person." Harrison (1994: 715) concurred, arguing that the provision of tourism training institutes enhances the self-image of tourism employees, "who might otherwise resent becoming a 'nation of bell hops and chambermaids'," especially in developing countries with a history of colonialism.

As stated earlier, Bali has a number of tourism training facilities in place, at both high-school and tertiary levels. The research demonstrated that training institutes such as BPLP enhance employability and facilitate employment that contributes to self-development. Creating a program to fund or subsidize the annual Rp700,000 student fee for BPLP would result in students from lower-income families being able to participate in the programs offered by the school.

As well as providing training for those who will be employed in the tourism sector, education on the impacts of tourism development should be provided to tourism administrators, policy makers and government tourism planners in developing countries so they can maximize the potential benefits of tourism. These influential groups are often uninformed about tourism, and where education is provided it is not adapted to the context and specific problems of that country.

Conclusion

Tourism in Bali, though still a relatively recent phenomenon, has developed to the point where it is actively promoted by the regional and national governments as an important economic sector, and has had an overall positive effect on those employed within that sector. Tourism employment in Bali is increasingly being seen as a desirable alternative to traditional occupations and as a provider of high incomes and high-status occupations. If people believe their work to be personally satisfying and of high status, "development" will be positively affected. An official policy of tourism employment promotion should result in greater investment in education which can benefit the community as a whole and not merely those seeking work in the tourism sector. Murphy (1985) wrote that tourism development can be positive if the needs of the community are placed before the goals of the tourism industry.

Much of the academic literature relating to tourism employment in developing countries has adopted a negative stance, focusing on the seasonal and servile nature, supposed low status, and high leakages associated with tourism employment. These analyses and criticisms are often based on assumptions which distort the overall picture. Seasonal aspects had a minimal impact on employment creation in Bali, and even where tourist arrivals reflected seasonal variations, employment opportunities were relatively unaffected. Tourism employment, whether part of the formal or informal sector, was generally perceived to have a high status and was not considered to be "servile" in nature. Thus guides may serve as testament that a new tourism employment option can quickly acquire a high status within the community, while at the same time providing individuals with an economically viable career.

Notes

1   At the time of this study (1991–3), US$1 was equal to Rp2,000, while the average monthly minimum wage in Bali was Rp84,000.
2   There are a number of tourism training institutions in Bali, some at the secondary school level and two at the post-secondary level. One of those at the post-secondary level is the Hotel and Tourism Training Institute of Bali (*Balai Pendidikan dan Latihan Pariwisata* or BPLP), established in 1978. BPLP is the main tourism training institute, training over 1,500 students a year. The institute has a number of different programs ranging in length from two weeks to four years, and provides training for a variety of positions, including those at hotels, restaurants, craft industries, middle management, guiding and entrepreneurship opportunities. Fees are Rp700,000 per year and include the cost of courses, uniforms and supplies. The institute has a successful job placement rate, with 80 percent of graduates employed in the tourism sector and 80–90 percent of those within hotels (Hanley 1989; Bagiarta 1993).
3   Official guides are those who are formally registered with the Bali Government. There are, however, many "unofficial" guides whose actual numbers have not yet been documented.
4   The *Bali Post* is the Bali provincial daily newspaper.
5   The 1992 population of both Sanur and Kuta was reported by Picard (1992) to be 15,000.

References

Akroun, M. (1992) "Cultural tourism and religious belief systems: an overview," in W. Nuryanti (ed.) *Universal Tourism: Enriching or Degrading Culture?*, Yogyakarta: Gadjah Mada University Press, 53–9.

Bagguley, P. (1987) "Flexibility, restructuring and gender: changing employment in Britain's hotels," cited in Hudson and Townsend (1992) "Employment and policy for local government," in P. Johnson and B. Thomas (eds) *Perspectives on Tourism Policy*, London: Mansell, 49–68.

Bagiarta, N. (1991) Director, BPLP; personal communication, May 9.

——(1993) personal communication, April.

Bali Government Tourism Office (1990) *Tourism in Bali*, Denpasar: Government Tourism Office.

——(1993) *Tourism in Bali*, Denpasar: Government Tourism Office.

Baum, T. (1993) "Human resources in tourism: an introduction," in T. Baum (ed.) *Human Resource Issues in International Tourism*, Oxford: Butterworth-Heinemann, 3–21.

Blanton, D. (1981) "Tourism training in developing countries: the social and cultural dimension," *Annals of Tourism Research*, 8, 1: 116–33.

Bromley, R. and Gerry, C. (1979) "Who are the casual poor?," in R. Bromley and C. Gerry (eds) *Casual Work and Poverty in Third World Cities*, Chichester: John Wiley and Sons, 3–23.

Cukier, J., Norris, J. and Wall, G. (1996) "The involvement of women in the tourism industry of Bali, Indonesia," *Journal of Development Studies*, 33, 2: 248–70.

Cukier-Snow, J. and Wall, G. (1993) "Tourism employment: perspectives from Bali," *Tourism Management*, 14, 3: 195–201.

——(1994) "Informal tourism employment: vendors in Bali, Indonesia," *Tourism Management*, 15, 6: 464–7.

Diamond, J. (1977) "Tourism's role in economic development: the case reexamined," *Economic Development and Cultural Change*, 25: 539–53.

Duffield, B. S. (1982) "Tourism: the measurement of economic and social impact," *Tourism Management*, 3, 4: 248–55.

Ebery, M. G. and Forbes, D. K. (1985) "The 'informal sector' in the Indonesian city: a review and case study," in G. H. Krause (ed.) *Urban Society in Southeast Asia: Social and Economic Issues*, Hong Kong: Asian Studies Monograph Series, 153–70.

Elkan, W. (1975) "The relation between tourism and employment in Kenya and Tanzania," *Journal of Development Studies*, 11, 2: 123–30.

Francillon, G. (1990) "The dilemma of tourism in Bali," in W. Beller, P. d'Ayala and P. Hein (eds) *Sustainable Development and Environmental Management of Small Islands*, Paris: UNESCO, 267–72.

Geertz, C. (1973) *The Interpretation of Cultures: Selected Essays*, New York: Basic Books.

Geertz, H. and Geertz, C. (1975) *Kinship in Bali*, Chicago IL: University of Chicago Press.

Geriya, I. W. (1991) UNUD; personal communication, October.

——(1993) personal communication, May 28.

Gertler, L. (1992) "Linkage between past, present and future," in W. Nuryanti (ed.) *Universal Tourism: Enriching or Degrading Culture?*, Yogyakarta: Gadjah Mada University Press, 11–18.

Gordon, A. (1978) "Some problems of analyzing class relations in Indonesia," *Journal of Contemporary Asia*, 4: 210–18.

Hanley, M. L. (1989) "Apprendre à sculpter une divinité éléphant en argile," *Développement Mondial*, 2, 6: 15.

Harrison, D. (1994) "Learning from the old south by the new south? The case of tourism," *Third World Quarterly*, 15, 4: 707–21.

Hudson, R. and Townsend, A. (1992) "Employment and policy for local government," in P. Johnson and B. Thomas (eds) *Perspectives on Tourism Policy*, London: Mansell, 49–68.

Hussey, A. (1989) "Tourism in a Balinese village," *Geographical Review*, 79, 3: 311–25.

Lindayati, R. and Nelson, G. (1995) "Land use change in Bali: a study of tourism development in Kelurahan Sanur," in S. Martopo and B. Mitchell (eds) *Bali: Balancing Environment, Economy and Culture*, Department of Geography Publication Series No. 44, Waterloo, Canada: University of Waterloo, 411–36.

Lovel, H. and Feuerstein, M. T. (1992) "Editorial introduction: after the carnival— tourism and community development," *Community Development Journal*, 27, 4: 335–52.

Mandra, I. M. (1993) Head of Operations, BTDC; personal communication, May 18.

Manuaba, A. (1991) UNUD; personal communication, May 7.

——(1993) personal communication, September.

Murphy, P. E. (1985) *Tourism: A Community Approach*, New York and London: Routledge.

Picard, M. (1990) "Cultural tourism in Bali: cultural performances as tourist attraction," in A. Kahin (ed.) *Indonesia*, Cornell NY: Cornell University Southeast Asia Program, 37–74.

——(1992) *Bali: Tourisme Culturel et Culture Touristique*, Paris: Éditions L'Harmattan.

——(1993) "Cultural tourism in Bali: national integration and regional differentiation," in M. Hitchcock, V. T. King and M. J. G. Parnwell (eds) *Tourism in Southeast Asia*, London and New York: Routledge, 71–98.

Pigram, J. J. (1990) "Sustainable tourism: policy considerations," *The Journal of Tourism Studies*, 1, 2: 2–9.

Pizam, A., Milman, A. and King, B. (1994) "The perceptions of tourism employees and their families toward tourism: a cross-cultural comparison," *Tourism Management*, 15, 1: 53–61.

Pongsapich, A. (1982) "Interplay of tradition and modernization in the fishing and tourist industries of Thailand," in G. B. Hainsworth (ed.) *Village-Level Modernization in Southeast Asia*, Vancouver, Canada: University of British Columbia Press, 335–56.

Rajotte, F. (1987) "Safari and beach resort tourism: the costs to Kenya," in S. G. Britton and W. C. Clarke (eds) *Ambiguous Alternative*, Suva: University of the South Pacific, 78–90.

Richter, C. (1986) "Tourism services," in O. Giarini (ed.) *The Emerging Service Economy*, Oxford: Pergamon, 213–44.

Sudiana, I. (1993) Personnel Manager, Bali Dynasty Resort; personal communication, May 18.

Sukabrata, I. W. (1993) Director of Human Resources, Bintang Bali Hotel; personal communication, May 20.

Turner, L. (1973) "Tourism: the most subversive industry," in L. Turner (ed.) *Multinational Companies and the Third World*, New York: Hill and Wang, 210–29.

UN Development Program (1992) *Draft Final Report: Strategies, Comprehensive Tourism Development Plan for Bali*, Volumes I, II and annexes, Denpasar: UNDP.

UNESCO (1976) "The effects of tourism on socio-cultural values," *Annals of Tourism Research*, 4, 2: 74–105.

Urry, J. (1990) *The Tourist Gaze: Leisure and Travel in Contemporary Societies*, London: Sage.

Vaughan, R. and Long, J. (1982) "Tourism as a generator of employment: a preliminary appraisal of the position in Great Britain," *Journal of Travel Research*, 21, 2: 27–31.

Vickers, A. (1989) *Bali: A Paradise Created*, Ringwood VIC, Australia: Penguin Books.

Wijaya, N. G. (1993) Kasi Pramuwisata dan Angukatan Wisata: Guide Committee; personal communication, May.

Wilkinson, P. F. and Pratiwi, W. (1995) "Gender and tourism in an Indonesian village," *Annals of Tourism Research*, 22, 2: 283–99.

Young, G. (1973) *Tourism: Blessing or Blight?*, Harmondsworth: Penguin Books.

5

REWRITING LANGUAGES OF GEOGRAPHY AND TOURISM

Cultural discourses of destinations, gender and tourism history in the Canadian Rockies[1]

Shelagh J. Squire

Drive through the Rockies and you'll enjoy a Canadian vacation classic. While relaxing in the comfort of your car or touring coach, a succession of picture postcard scenes rolls past like your own personal travelogue.

(Palmer 1996: 12)

The type of people who came [to the Rockies] were more the wealthy people because they were the only people who could afford . . . to stay at hotels and hire horses.

(Tocher 1977)

This issue . . . is about a declining ecosystem, and our inability to come to terms with what needs to be done to restore it. . . . I foresee a terrible day when the only bears left in Banff National Park will be found in cash register drawers, stamped on the face of two-dollar coins.

(Legault 1996: 15)

These extracts capture different facets of the tourist destination that was and is the Canadian Rocky Mountains. The scenic beauties of Canada's western mountains are the currency for international tourist recognition. Correspondingly, however, the mountains as a destination are also a social construction. The landscape preferences and values for wilderness to which they appeal reflect cultural influences originating in European Romanticism. As tourist space, the Rockies have been shaped by differences of race, class, gender and a dialectic between work and leisure. And, not least, the mountains have been a milieu where notions of limits to growth and ideas about sustainable tourism have collided with the pressures of over-development and economic globalization.

It is only recently that geographers have begun to invoke critical social theory to enhance understandings of tourism activities and processes, and to integrate thereby well-established economic and growth-related concerns with other, more qualitative and interpretive dimensions of the touristic experience (Britton 1991; Hughes 1992; Squire 1994a; Crang 1996). Yet as the languages of geography and tourism are rewritten slowly, ideas about tourism as a social and cultural construct remain somewhat distant from the cultural critiques that have already altered significantly the ways in which various other disciplinary sub-fields are being conceptualized and defined.

This chapter considers aspects of the changing languages of geography and tourism, focusing in particular on how recent reformulations of culture, society and space necessitate a re-examination of discourses of destinations.[2] First, a brief overview of the extant research realm of tourism geography will be presented. Then the concept of destination itself will be assessed, paying particular attention to how certain theoretical constructs and related cultural critiques impact on the ways that destinations may be interpreted. Finally, conjoining theory and method, a case example of research in progress, about gender and tourism history in the Canadian Rocky Mountains, will be examined to highlight the way new languages of geography and tourism may operate in research practice.

Writing the tourist landscape is as much an empirical as it is a theoretical task. It is also a task that extends beyond conventional disciplinary boundaries. Tourism studies have long been a multidisciplinary activity. By way of conclusion, connections between disciplinary and interdisciplinary ways of knowing will be explored, and suggestions made for the possible implications for tourism geography of some emerging and reconfigured intellectual terrains.

Geography and tourism studies

Tourism and geographical appraisals thereof have been summarized elsewhere (Britton 1979; Pearce 1979; Smith and Mitchell 1990; Mitchell and Murphy 1991; Squire 1994a). And while this chapter's focus is on the changing languages of tourism geography, it must also be emphasized that these languages are to some extent allied with parallel dialogues in leisure studies. For example, there is a large amount of literature pertaining to aspects of leisure and recreation (Henderson *et al.* 1989; Jackson and Burton 1989; Rojek 1990; Warren 1993) and also to connections between tourism, recreation and leisure (Mieczkowski 1981; Colton 1987; Fedler 1987; Mannell and Iso-Ahola 1987; Butler 1989). To some extent these issues intersect with concerns addressed here.

What, then, are the key issues with which tourism geographers have engaged? This research area has always had a strong spatial and economic focus. Thus much work has examined tourism and tourist-related activities in relation to land use and infrastructure development, multiplier effects, visitor flows and resort cycles of evolution (selected examples include Wolfe 1951; Hills and Lundgren 1977; Butler 1980; Mansfeld 1990; Agarwal 1994; Burton 1994; Timothy and Butler 1995). Many tourism texts echo these perspectives, defining tourism implicitly as an industry occurring in particular spatial contexts (Mathieson and Wall 1982; Pearce 1989; Smith, S. 1989). Complementing these economically oriented approaches, however, are also a range of other traditions, demonstrating aptly the diversity of geographic contributions to the field.

Most recently, geographers have been active in studies of tourism and sustainability. While suggesting definitions of sustainable tourism development (Butler 1993), such research has also significantly advanced empirical understandings of tourism sustainability and limits to growth (selected examples include Bramwell and Lane 1993; Hughes 1995b; Manning and Dougherty 1995; Wall 1996). Finally, some attention has focused on aspects of tourism history (Marsh 1982; Butler 1985; Butler and Wall 1985; Towner 1988, 1994), marketing and tourism images (Dilley 1986; Ashworth and Goodall 1990; Goss 1993), and tourism-related urban heritage (Ashworth and Tunbridge 1990; Chang *et al.* 1996; Tunbridge and Ashworth 1996; Waitt and McGuirk 1996).

In sketching these parameters of tourism geography, a broad brush has been intentionally used. This summary is neither exhaustive nor especially inclusive. Rather, themes and issues cited provide a background for the discussion that follows: one that seeks to link tourism research with contemporary theoretical constructs about culture, society and space.

Rewriting the languages of geography and tourism necessitates much more disciplinary, and indeed interdisciplinary, dialogue. Thus while not denying the importance or continued significance of spatial and economic questions, an argument is set forth here for a concomitant assessment of other ways of seeing and knowing tourism processes; an assessment that is to some extent premised on revisiting the concept of destination itself.

Discourses of destinations

At a superficial level, the term "destination" would seem relatively nonproblematic. After all, much tourism research has considered impacts of tourism on people and places. What is such research, then, but a study of different aspects of destinations? In a similar vein, destinations are also the focus for a variety of tourism-related promotional activities, elements of which have been studied widely. For the intent of this project, destinations are not merely a leitmotif for geographic place. Rather, they are also social and cultural constructions whose meanings and values are negotiated and redefined by diverse people, and mediated by factors often related only tangentially to a particular tourist setting.

Theory

Cultural critiques and attendant theoretical frameworks have transformed the nature of intellectual inquiry in many parts of this discipline. For example, ideas about cultural production, consumption and representation continue to have enormous currency in human geography generally, and social and cultural geography in particular (Gregson 1995). Not least, such ideas have suggested a multitude of questions, both theoretical and applied, about how landscapes, places and human–environmental relationships may be conceptualized (for a review of relevant literature see Duncan 1993, 1994, 1995; Gregson 1995; Matless 1995, 1996). To date, however, there have been few points of contact, at least in the published geographical literature, between such researchers and tourism specialists.

In social and cultural geography, edited collections devote little significant attention to tourism topics (as a case in point see Kobayashi and Mackenzie 1989; Gregory and Walford 1989; Anderson and Gale 1992; Duncan and Barnes 1992; Duncan and Ley 1993). And even when tourism issues might be deemed central, as in studies of past travelers and travel writers, tourism scholarship has gone virtually uncited (Blunt 1994; Blunt and Rose 1994; McEwan 1996). Correspondingly, though, it is only very recently that cultural theories and perspectives have even been explored by geographers engaged in tourism research.

In 1991, Hughes lamented what he then called the "pre-social state" of tourism geography, "unaware of the most basic contests of knowledge arising from the structure agency debate" (Hughes 1991: 226). While the cultural studies literature, broadly defined, suggests diverse opportunities for tourism scholarship, most of such assessments have taken place outside geography. For instance, researchers have used semiotic and cultural materialist approaches to assess tourism destinations and tourist practices (Bennett 1986; Adler 1989; Cohen 1989; Evans-Pritchard 1989; MacCannell 1989). Similarly, much provocative work has attempted to forge links between aspects of tourism, culture and society (Cohen 1993; Graburn 1995; Dann 1996). And in this sense, Urry's contributions are especially important, not least his recent assessment of consumption (Urry 1990, 1995).

Since Hughes' comment, however, there have been some encouraging signs that tourism geography has drawn closer to consideration of the cultural issues that continue to impact strongly on research questions elsewhere in our discipline. Various writers have focused on aspects of the production and consumption of tourism sites and activities and have theorized tourism, either explicitly or implicitly, as a form of cultural communication (e.g. Shaw and Williams 1994; Squire 1994a; Crang 1996). Questions have also been raised about touristic meanings and values (Squire 1994b; Hughes 1995a; Waitt and McGuirk 1996), and links between tourism, critical economy and postmodernity (Shields 1990; Britton 1991; Hughes 1992, 1995b; Oakes 1993). And, from multiple disciplinary perspectives, geography included, the significance of race and gender in tourism contexts is also being explored.

In this latter sense, and for the most part, scholarly appraisals of tourism have tended to take one of two forms (on a related point see Bella 1989). First, they are usually conceptualized within masculine frameworks. As represented, the dominant tourist experience is that of a white male traveler or less commonly, worker (see for example Towner 1984; Marsh 1985). Second, such accounts are frequently gender- and/or racially neutral. From this perspective, individual tourists are described via analytic categories or through macro-scale assumptions effectively removed from the actual touristic experience (a well-known example is MacCannell 1976; for a general critique see Veijola and Jokinen 1994). Recent attention has focused on the manner in which constructs like race and gender, and to some extent sexuality, shape the ways that tourist settings may be created, experienced and interpreted over space and time (Kinnaird and Hall 1994; Norris and Wall 1994; Squire 1994c; Richer 1995; Swain 1995; Holcomb and Luongo 1996).

Consideration of issues such as these is fostering new languages of tourism geography. In traditional definitions of tourism, such as travel and absence from home (Holloway 1987), destinations are to some extent categories for statistical analysis, for instance of economic impacts or place perceptions. In a second sense—one that is the focus of this chapter—destinations are also sites where cultural ideologies, expectations and traditions are played out and are contested by different social actors and interest groups. Thus as both geographic place and theoretical construct, destinations are cultural texts that invite a multitude of readings and interpretations. While the landscape as text metaphor has been used widely in human geography to assess critically diverse kinds of human–environmental interaction (selected examples include Cosgrove and Daniels 1988; Duncan and Duncan 1988; Duncan and Barnes 1992), tourism landscapes have remained almost immune from theoretically informed textual analyses. Yet such textual landscape readings may help to cast tourism more fully as leisure and labor, and also serve to focus greater scrutiny on the inner workings of tourism practices and processes.

Methodology and methods

As noted, theoretical conceptions of tourism and destinations increasingly transcend disciplinary boundaries. At a macro-level, the implications of these shifting intellectual terrains and geography's role therein, are highlighted in the concluding part of this chapter. In the present context, though, this breaking-down of disciplinary barriers and the changing foci for tourism geography that it implies, also raises methodological concerns.

Across the social sciences, research practice has in recent years been a focus of increased scrutiny. Much of this reflection has been concerned with qualitative approaches: as related to both quantitative methods of data analysis, and as a way of knowing that raises new questions about bias, objectivity and data interpretation. In human geography, for example, the collection by Eyles and Smith (1988) was one of the first comprehensive statements about qualitative methodology and how qualitative methods might be applied empirically. More recently, feminist geographers have demonstrated particular concern for methodological issues, not least in emphasizing reflexivity, positionality and research ethics (see for example Dyck 1993; Moss 1993; Rose 1993; Nast 1994).

As related to tourism studies, questions of appropriate methods and methodologies have also been addressed (see Towner 1984; Cohen 1988; Dann *et al.* 1988; Smith, S. 1989; Smith and Mitchell 1990; Squire 1994a). While it lies beyond the scope of this chapter to discuss methodology in any detail, I would stress that theory and method must to some extent be considered together. Thus fresh theoretical perspectives necessitate correspondingly reappraisals of empirical research practice. In geographic contexts, the traditional emphasis on spatial and economic aspects of tourism development has meant the dominance of quantitative and statistically oriented work. Such approaches, although important, can contribute only a partial understanding of tourism processes. In tourism geography there is thus tremendous scope to explore further how the kind of qualitative research being carried out fruitfully elsewhere in our discipline, and across the social sciences more generally, might be integrated further into our analyses.

Examining social and cultural questions through tourism necessitates collection of a range of data. Open-ended interview questions and focus-group strategies, for instance, can generate important insights about the nature of tourism as experienced by different players within touristic systems. Yet in most tourism research, and as noted earlier from a theoretical perspective, it is rare that individual voices and/or interpretations are heard. Rather, in terms of methodology too, the tourism experience is mediated most frequently by the researcher and attendant research agendas privileging the collective over the particular.

If, however, individual voices are highlighted, questions of data analysis also emerge. Strauss (1987) offers guidelines for qualitative textual interpretation that may be adapted readily into tourism settings. The literature on discourse analysis (e.g. Fairclough 1992) offers yet another medium for data appraisal. And there are a growing number of computer software packages designed explicitly to manage and assist with qualitative data interpretation.

If tourism sites and activities are produced in different ways and for different purposes, and are also interpreted through perceptual filters shaped amongst other things by race, social class, gender and sexual orientation, then empirical practice must shift accordingly. Since tourist destinations carry multiple meanings, qualitative and quantitative approaches used together can highlight both the multiplicity of those meanings as well as the interconnections between them. Cultural discourses of destinations embrace both theoretical and methodological concerns. Yet how may these new languages of geography and tourism be adapted into particular research contexts?

Gender and tourism history in the Canadian Rockies

As noted above, it is only recently that gender issues have been drawn into discourses of tourism and destinations. Thus in relation to historiographies of tourism, Towner (1994: 725) explains: "Women have generally remained hidden from tourism history, whether as tourists . . . or as workers . . . yet their significance was clearly very great." The history of tourism development in Canada's Rocky Mountains is a case in point. Most accounts of regional tourism emphasize men's contributions as explorers, railway builders, financiers and tourist guides and outfitters (Fraser 1969; Hart 1979; McKee and Klassen 1983; Marsh 1985). Women did, however, play various roles in these past tourism processes (Dixon 1985; Smith, C. 1989; Squire 1995a). And in terms of labor as well as leisure, women's contributions enhance understandings of a destination that to date has tended to be represented as either gender-neutral or has been constructed predominantly through masculine lenses.

Drawn from research in progress, the following case-study is supported theoretically by ideas about cultural production, consumption and representation (on a related point see Johnson 1986; Squire 1994a, 1994b). And, following Strauss (1987), the research is based upon a detailed textual analysis of women's diaries, letters and interviews, as well as of advertising material produced by the Canadian Pacific Railway (CPR), a major player in the early tourism industry.

Producing a destination

In this example, questions of gender and tourism history are framed necessarily within the larger context of regional tourism development. In the Canadian Rockies, tourism dates from the mid-1880s and is associated primarily with completion of the mountain portion of the CPR's transcontinental rail line in southern Alberta and British Columbia. Promoting mountain scenery to tourists was a means of deriving revenue from an otherwise costly construction initiative (MacBeth 1924; Fraser 1969; Hart 1983; McKee and Klassen 1983). Hence the railway invested heavily in tourist infrastructure development. Luxury hotels were built at Banff, Lake Louise and in the Selkirk Mountains (Glacier House) of British Columbia. Tourist trails soon dotted the mountains near Banff town-site and, linked with railway hotels throughout the mountain region, guiding services for hiking, trail riding and mountaineering were also provided.

To advertise the region, the CPR produced pamphlets, guidebooks, posters and related ephemera. Such materials extolled not only wilderness scenery but also the amenities, both culinary and recreational, available at the mountain resorts. This publicity drew the Rockies into romantic discourses of alpine scenery (fostered in part by extant public interest in the Swiss Alps) as well as into Canadian Pacific's corporate culture. Even though the Canadian government had passed legislation leading to the formation of several National Parks and protected forest reserves in this western region (see also Bella 1987), numerous advertisements described the mountains as the "Canadian Pacific Rockies" (Hart 1983). Targeting the affluent social elite, this advertising was circulated widely throughout eastern Canada, parts of the United States and western Europe.

This phase of mountain tourism effectively ended in 1939 with the outbreak of the Second World War in Europe. Although certainly affected by conditions of economic depression in the 1930s, regional tourism until 1939 took much the same form as it had since the 1880s. Despite the emergence of more hotels, guest houses and recreational facilities, most visitors still arrived in the area by train, the CPR controlled much of the tourist infrastructure, and wealthy, often international visitors constituted the primary tourist market.

Women in early mountain tourism

The prewar period, and specifically how women acted as both producers and consumers of this tourist place, are the focus here. Much like men, women played multiple and multifaceted roles in early mountain tourism. Images of women also featured prominently in regional publicity. Considered together, then, women's experiences and representations thereof, contribute to an alternative and specifically gendered discourse of this destination and the patterns of tourism development that it encompasses. Drawn from a larger research project, the discussion that follows is of necessity a cursory sketch. Not so much a study of the destination itself, this example is illustrative rather of how aspects of the changing languages of geography and tourism outlined in the first part of this chapter might function empirically.

Travelers

In the first instance, and making manifest theoretical ideas about tourism as cultural production, consumption and communication, many women were enthusiastic travelers, sometimes engaging directly in exploration, geographic research and mountaineering. One of the earliest and best known of such women was Mary Schäffer Warren, a Quaker from Pennsylvania who discovered the Rockies through a family holiday and later accompanied her first husband on several botanical expeditions. After her husband's death, she contracted with local guides to take her further into the mountains, and in 1907 and 1908 she and a female traveling companion financed and participated in two extended exploratory ventures. On a trip in 1911, she was accompanied by her sister-in-law (Adams 1908; Gowan 1957; Hart 1980; Smith, C. 1989; Squire 1995a).

Mary Schäffer Warren wrote about her mountain travels, and these writings were in turn published and read extensively in North America and Britain (Schäffer 1907, 1908, 1910, 1911). Hence, while she was first a consumer of the touristic experience, staying at the CPR resorts when not camping out, and relying upon the services of regional guides and wilderness outfitters, she was also in effect contributing to the production of the destination itself. Subsequent visitors and authors in their own right referred to her work (e.g. Kipling 1908; Cran 1911: 183), and her way of seeing Canada's western mountains presumably influenced others' interpretations too.

In the case of the Canadian Rockies, there are numerous examples of such aspects of gendered tourism history and, in relation to theoretical constructs outlined earlier, of how tourist activities were embedded within a much larger social and cultural milieu. Mary Vaux, another American visitor, pursued studies in glaciation. Together with family members, she pioneered some of the first documented research on Canada's glaciers (Cavell 1983). In her case, scientific fieldwork was combined with an extended summer holiday at one or more of the mountain resorts. Circulating her research results via both public lectures and geographical journals, Vaux, like Schäffer Warren, was both consumer and producer of particular elements of this destination space. Another traveler was Carolyn Hinman, first a tourist and subsequently a regional travel promoter. From her home in New Jersey, she came to the Rockies initially to participate in Alpine Club of Canada summer camps. Some years later she launched her own travel business and through her "Off the Beaten Track" tours brought many parties of young American women on escorted (and closely chaperoned) mountain wilderness excursions (Hinman 1915–1960; Smith 1989).

Much more could be written about these and other women who traveled in the mountain region. The texts that they left behind, both published and unpublished, are rich in details of early mountain tourism generally, and in particular as experienced by certain wealthy and well-educated individuals. Their observations about wilderness activities, hotel life and scenery (of which aboriginal peoples formed part), offer fresh insights into regional tourism history and the intersections between race, class and gender embedded therein. Not least, their accounts emphasize different facets of the tourist experience than do those written by comparable male raconteurs. For example, women diarists paid considerable attention to domestic aspects of tourism, something about which men were largely silent.

Gender, however, is not an all-encompassing construct. Thus in rewriting languages of geography and tourism, and in gendering tourist destinations past and present, it is important to account for differences not only between but also amongst men and women. In tourism contexts and, as noted briefly above, such differences are shaped in part by a dialectic between work and leisure.

Workers

The production and maintenance of the Canadian Rocky Mountains as a tourist destination required the efforts of a sizeable labor force, many of whom were women working for railway hotels and ancillary tourism services. In the National Parks and related forestry organizations, wives of park wardens and forest rangers contributed ample unpaid labor in activities related tangentially to regional tourism development. Except as a statistical category, tourism geographers have been loathe to assess "work" and its interconnections with leisure. Yet, as a strand within gendered tourism places and as part of the social context within which tourism is enmeshed, "work" merits much further scrutiny (although see Kinnaird and Hall 1994).

In terms of textual records, information about ordinary tourism workers is usually more limited than it is for those who traveled or who worked in tourism in "professional" capacities. This case example is no exception. Kate Reed who fashioned the decor of many early CPR hotels, those in the mountains included, kept detailed diaries of her experiences. Her husband, however, was a senior CPR official and she herself was from a prominent Canadian family (Anonymous 1911; Reed 1856–1928). Although these diaries offer insights into particular hotels and tourist activities in the mountains, they are in no way representative of an ordinary woman's working life.

From various sources, though, a partial picture emerges, one that underscores differences between individual women in terms of how the tourism context was then experienced. Alice Harding came from London, England, to work as a chambermaid at Glacier House Hotel. Describing her working life, she noted that at certain times "you did everything—you knew what was to be done" (Harding 1970). Yet she also recalled that she was sometimes able to pursue climbing and hiking activities and to participate in the dances and social events organized by hotel staff.

But for others the busy summer season meant a much different working experience. Ivy Paris toiled in a Banff restaurant in the 1930s and recalled "We used to work eighteen hours one day and nine the next in the tea room. We worked darn hard" (Paris 1984: 38). The restaurant owner, also a woman, "used to come to work at noon-time and [would] be there until two o'clock in the morning" (ibid.). In addition to tea-room and restaurant operations, other female-headed businesses catering to the tourist trade included souvenir and craft shops, news kiosks and tourist accommodations. A small number of women were even involved in tourist guiding and outfitting activities, although usually as part of an existing family business. And, in these different contexts, some evidence suggests that women were engaged in many of these activities either concurrently or at different times throughout tourist seasons (Unwin 1970).

Finally, the experiences of wives of park wardens and forest rangers offer yet another discourse of this destination. Ethelwyn Alford was the wife of a forest ranger, and in 1924, at the age of fifty-two, she accompanied her husband on his summer trail inspection duties. Her extant diary offers a particular view of regional development, one that is again distinct from more common touristic interpretations (Alford 1924). As an expatriate Englishwoman, Ethelwyn was to some extent a consumer, writing about mountain scenery in ways comparable to numerous travel writers of the period. Yet, plagued by ill health and burdened with arduous domestic chores, her experiences differed sharply from those of the leisured tourist parties that she and her husband encountered. While such groups occupied themselves with photography, botanical pursuits and "staging original entertainments out of doors" (Alford 1924: 8 August), she was a member of a working wilderness party, engaged in a tedious (and for her unpaid) round of cooking, washing and driving packhorses. From this perspective, then, her narrative highlights crucial differences between leisure and labor, and between tourist imagery and working reality. As she noted, "I never want to see a western movie again now that I have gone in for the real thing even this much" (Alford 1924: 8 July).

Such accounts of "work" are multifaceted, drawing attention to differences amongst women workers, but also underscoring women's contributions to tourism development historically, in both paid and unpaid capacities. Furthermore, such recollections raise important questions about how aspects of leisure and labor may have been intertwined for certain working women. In its multiple manifestations, then, and as related to meanings, values and culture, "work" necessitates further analysis.

Representations

The third strand in this gendered discourse is that of the images of women used regularly in CPR advertising. In historical contexts, the texts of tourism promotion remain somewhat unstudied (although see Hart 1983). Yet these materials can offer crucial insights into the values and ideologies that shaped destination development and that may also have strongly influenced how such destinations were perceived.

In CPR advertisements before 1939, images of women are far more prominent than are comparable images of men.[3] Royal Canadian Mounted Police (RCMP) officers and aboriginal men appear frequently, but if not supporting women (often as romantic partners) both male visitors and workers are scarce. By contrast and in highlighting recreational amenities at railway resorts, CPR advertising chronicles both aspects of women's multiple roles in the early mountain tourism industry, as well as the extant gender constructs within which they were enmeshed.

At that time, railway publicity made extensive use of female imagery to "decorate" and/or "enhance" both wilderness scenery and hotel facilities and attractions. In some advertisements women appeared as active mountaineers, hikers and trail-riders. More commonly, though, female models were depicted looking at scenery (often through hotel windows) or in provocative swimsuit poses. And in a background role, women also featured regularly as domestic servants.

While some of these advertisements captured elements of women's experiences in the mountain region, they also helped to associate the mountain landscape with a particular version of female beauty. Without exception, visitors were depicted as young, white, able-bodied and either single or newly married. And in certain cases, the written text accompanying illustrated advertisements differentiated parts of the tourist experience from what was then defined as women's primary social role. As one example from the 1920s queried:

> Where have they been all day, these butterfly ladies who flutter around the hotel at night in Paris gowns? Riding the trails and roaming the forests in strictly utilitarian togs. After sundown they *revert to type*, chattering over Banff's smart frocks, whispering over newly met "hims," in this purely feminine retreat. Women especially delight in Banff's refinements of service, its exquisite linens and china. . . . Another purely feminine joy is the beauty parlour where some of the smartest women in the world are skillfully attended.
> (Canadian Pacific Railway 1929: 15, emphasis added)

In this and other advertisements, heterosexual romance was featured and the Rockies were promoted frequently as a honeymoon destination. Directed towards women in particular, one such publication urged its readers to forge "two outstanding memories . . . your wedding day and the day you arrive at Banff" (Canadian Pacific Railway 1936: 3).

These texts send strongly gendered messages, and in also raising questions about identity, sexuality, race and social class amongst others, they constitute an important discursive element within the production and consumption of this particular destination space. Such issues are implied in elements of the new languages of geography and tourism described earlier. They have, however, only been the focus of limited empirical attention.

Synthesis and conclusions

As a process, tourism reflects the social and cultural contexts within which it is created, carried out and interpreted. As noted, the case example discussed in this chapter is neither exhaustive nor inclusive, but highlights briefly how a destination can be socially and culturally constructed, in this instance as mediated by gender influences and in particular women's roles. This research raises both theoretical and methodological questions. For example, what analytical frameworks account most fruitfully for the multiple cultures of tourism associated with particular tourist destinations? How are theoretical notions about cultural meanings, values and transformation best conjoined with empirical practice? How do race and class intersect with gender to mediate cultural and cross-cultural discourses of destinations? And not least, how do leisure, labor and particular power relationships interface to create and change a destination's profile over space and time?

Questions such as these extend far beyond the case example cited here. They impact moreover upon understandings of tourism past *and* present and are integral to the rewritten languages of geography and tourism described in the first part of this chapter. A range of recent writings has signaled new directions for tourism studies generally and tourism geography particularly. And re-theorized notions of culture, society and space and their incorporation into tourism contexts necessitate correspondingly a re-evaluation of methods and methodologies. In linking theory and method, however, questions about people and how they interact with destinations become singularly important. Invoking new languages to write the tourist landscape requires asking new questions about visitors and tourism workers, both individually and in terms of group contexts and interactions.

The case-study research cited here draws on the voices and images of actual women, and on representations of femininity and feminine behavior fueled by then-extant ideologies about gender and women's roles. This example therefore celebrates diversity in a way that has gone virtually unrecognized in both tourism studies and tourism geography. From a different vantage point, Lehr (1996) advances a similar argument as he notes:

> In our eagerness to reduce a varied, complex, and often chaotic process into neat and orderly paragraphs for respectable academic journals, there is a danger that those who write of the historical geography of Canadian settlement may lose sight of the individuals who helped to create the patterns described. If we ignore the range of diversity within broad patterns of group behavior we risk abandoning the essential dimension of the settlement process.
>
> (Lehr 1996: 107)

So, too, in writing about tourism have "varied, complex, and often chaotic process[es]" been subsumed within macro patterns and large-scale analytic categories. And, while group behavior is undeniably important, the new languages of geography and tourism outlined in this chapter should, in empirical practice, also result in increased attention to the particular individuals *within* groups who have shaped tourist landscapes past and present.

Ultimately, these emerging and reconfigured scholarly terrains intersect with both disciplinary and interdisciplinary ways of knowing. Much has been written about disciplinary and interdisciplinary knowledges and their connections at ideological, curricular and administrative levels (Swoboda 1979; Newell 1990; Stember 1991; Klein 1993; Nissani 1995; Squire 1995b). And while tourism studies have always been to some extent multidisciplinary, recent culturally inspired questions and critiques suggest exciting possibilities for greater interdisciplinarity in tourism research (see also Przeclawski 1993).

Discourses of destinations can only be understood fully from multiple perspectives and through multiple disciplinary lenses. The study of destinations is thus an inherently interdisciplinary activity. Yet interdisciplinarity depends in part upon disciplinary research, and effective interdisciplinary scholarship demands the thoughtful and critical integration of two or more disciplinary perspectives. How can disciplines be combined formally to address a range of tourism processes and issues most fruitfully? As integrative disciplinarians, geographers are well positioned to contribute significantly to such a question. And as disciplinary boundaries throughout the humanities and social sciences blur, the new languages of geography and tourism, and indeed the tourist landscape itself, continue to be reconfigured from multiple and multifaceted perspectives.

Notes

1   I thank the Central Research Fund, Carleton University, and the Central Research Fund, University of Alberta, for support of some of the research upon which this chapter is based. I also acknowledge gratefully the research assistance of staff at the Archives of the Whyte Museum of the Canadian Rockies (Banff, Alberta), the Corporate Archives of the Canadian Pacific Railway (Montreal, Quebec), Archives of the Glenbow Museum (Calgary, Alberta) and the Archives of the McCord Museum (Montreal).

2   In this context, "language" refers to the particular themes and issues that have shaped geographic examinations of tourism topics. Dann (1996) uses the term "language" somewhat differently, offering a fascinating sociolinguistic perspective on the language of tourism itself and the implications of this language for both tourists and tourism professionals.

3   A variety of tourist brochures and related advertising ephemera were consulted in collections held by the Corporate Archives of the Canadian Pacific Railway company, Montreal. Reproduction costs prevent inclusion of any of these advertisements here. For a sample of relevant materials see Hart (1983).

References

Adams, M. (1908) "Unpublished travel diary," Mary Schäffer Warren Fonds, Archives of the Whyte Museum of the Canadian Rockies, Banff, Alberta, M79/11.

Adler, J. (1989) "Origins of sightseeing," *Annals of Tourism Research*, 16, 1: 7–29.

Agarwal, S. (1994) "The resort cycle revisited: implications for resorts," in C. P. Cooper and A. Lockwood (eds) *Progress in Tourism, Recreation and Hospitality Management*, vol. 5, Toronto, Ontario: Wiley, 194–208.

Alford, E. (1924) "With the forest rangers in the Rockies," unpublished diary, Ethlewyn Alford Fonds, Archives of the Whyte Museum of the Canadian Rockies, Banff, Alberta, M488.

Anderson, K. and Gale, F. (eds) (1992) *Inventing Places: Studies in Cultural Geography*, Melbourne: Longman Cheshire.

Anonymous (1911) "What one woman earns," *Canadian Courier*, 4, 11: unpaginated, in Reed Family Papers, Hayter Reed Papers, Box 3, Folder 5, Archives of the McCord Museum, Montreal, Quebec.

Ashworth, G. J. and Goodall, B. (eds) (1990) *Marketing Tourism Places*, New York: Routledge.

Ashworth, G. J. and Tunbridge, J. E. (1990) *The Tourist-Historic City*, London: Belhaven.

Bella, L. (1987) *Parks for Profit*, Montreal: Harvest House.

——(1989) "Women and leisure: beyond androcentrism," in E. L. Jackson and T. L. Burton (eds) *Understanding Leisure and Recreation: Mapping the Past, Charting the Future*, State College, Pennsylvania: Venture Publishing, 151–79.

Bennett, T. (1986) "Hegemony, ideology, pleasure: Blackpool," in T. Bennett, C. Mercer and J. Woollacott (eds) *Popular Culture and Social Relations*, Milton Keynes: Open University Press, 135–54.

Blunt, A. (1994) *Travel, Gender and Imperialism: Mary Kingsley and West Africa*, New York: Guilford Press.

Blunt, A. and Rose, G. (eds) (1994) *Writing Women and Space: Colonial and Postcolonial Geographies*, New York: Guilford Press.

Bramwell, B. and Lane, B. (1993) "Sustainable tourism: an evolving global approach," *Journal of Sustainable Tourism*, 1, 1: 1–5.

Britton, R. (1979) "Some notes on the geography of tourism," *The Canadian Geographer*, 23, 3: 276–82.

Britton, S. (1991) "Tourism, capital and place: towards a critical geography of tourism," *Environment and Planning D: Society and Space*, 9, 4: 451–78.

Burton, R. C. J. (1994) "Geographical patterns of tourism in Europe," in C. P. Cooper and A. Lockwood (eds) *Progress in Tourism, Recreation and Hospitality Management*, vol. 5, Toronto, Ontario: Wiley, 3–25.

Butler, R. W. (1980) "The concept of a tourist area cycle of evolution: implications for management of resources," *The Canadian Geographer*, 24, 1: 5–12.

——(1985) "Evolution of tourism in the Scottish Highlands," *Annals of Tourism Research*, 12, 3: 371–91.

——(1989) "Tourism and tourism research," in E. L. Jackson and T. L. Burton (eds) *Understanding Leisure and Recreation*, State College, Pennsylvania: Venture Publishing, 567–95.

——(1993) "Tourism: an evolutionary perspective," in J. G. Nelson, R. W. Butler and G. Wall (eds) *Tourism and Sustainable Development: Monitoring, Planning, Managing*, Waterloo, Ontario: Heritage Resources Centre University of Waterloo, 27–43.

Butler, R. W and Wall, G. (1985) "Introduction: themes in research on the evolution of tourism," *Annals of Tourism Research*, 12, 3: 287–96.

Canadian Pacific Railway (1929) *Banff: Banff Springs Hotel, Canadian Rockies*, Montreal: Canadian Pacific.

——(1936) *Castle in the Air*, Montreal: Canadian Pacific.

Cavell, E. (1983) *Legacy in Ice: The Vaux Family and the Canadian Alps*, Banff, Alberta: The Whyte Foundation.

Chang, T. C., Milne, S., Fallon, D. and Pohlmanh, C. (1996) "Urban heritage tourism: the global-local nexus," *Annals of Tourism Research*, 23, 2: 284–305.

Cohen, E. (1988) "Traditions in the qualitative sociology of tourism," *Annals of Tourism Research*, 15, 1: 29–46.

——(1989) "Primitive and remote: hill tribe trekking in Thailand," *Annals of Tourism Research*, 16, 1: 30–61.

——(1993) "The study of touristic images of native people: mitigating the stereotype of a stereotype," in D. G. Pearce and R. W. Butler (eds) *Tourism Research: Critiques and Challenges*, New York: Routledge, 36–69.

Colton, C. W. (1987) "Leisure, recreation, tourism: a symbolic interactionism view," *Annals of Tourism Research*, 14, 3: 345–60.

Cosgrove, D. and Daniels, S. (eds) (1988) *The Iconography of Landscape*, Cambridge: Cambridge University Press.

Cran, M. D. (1911) *A Woman in Canada*, London: W. J. Ham-Smith.

Crang, M. (1996) "Magic kingdom or a quixotic quest for authenticity?" *Annals of Tourism Research*, 23, 2: 415–31.

Dann, G. (1996) *The Language of Tourism: A Sociolinguistic Perspective*, New York: CAB International.

Dann, G., Nash, D. and Pearce, P. (1988) "Methodology in tourism research," *Annals of Tourism Research*, 15, 1: 1–28.

Dilley, R. S. (1986) "Tourist brochures and tourist images," *The Canadian Geographer*, 30, 1: 59–65.

Dixon, A. (1985) *Silent Partners: Wives of National Park Wardens*, Pincher Creek, Alberta: Dixon and Dixon.

Duncan, J. S. (1993) "Landscapes of the self/landscapes of the other(s): cultural geography 1991–92," *Progress in Human Geography*, 17, 3: 367–77.

——(1994) "The politics of landscape and nature, 1992–93," *Progress in Human Geography*, 18, 3: 361–70.

——(1995) "Landscape geography, 1993–94," *Progress in Human Geography*, 19, 3: 414–22.

Duncan, J. S. and Barnes, T. J. (eds) (1992) *Writing Worlds: Discourse, Text, and Metaphor in the Representation of Landscape*, London: Routledge.

Duncan, J. S. and Duncan, N. (1988) "Re-reading the landscape," *Environment and Planning D: Society and Space*, 6, 2: 117–26.

Duncan, J. S. and Ley, D. (eds) (1993) *Place/Culture/Representation*, New York: Routledge.

Dyck, I. (1993) "Ethnography: a feminist method?," *The Canadian Geographer*, 37, 1: 52–7.

Evans-Pritchard, D. (1989) "How 'they' see 'us': Native American images of tourists," *Annals of Tourism Research*, 16, 1: 89–105.

Eyles, J. and Smith, D. M. (eds) (1988) *Qualitative Methods in Human Geography*, Cambridge: Polity Press.

Fairclough, N. (1992) *Discourse and Social Change*, Cambridge: Polity Press.

Fedler, A. J. (1987) "Are leisure, recreation and tourism interrelated?" *Annals of Tourism Research*, 14, 3: 311–13.

Fraser, E. (1969) *The Canadian Rockies: Early Travels and Explorations*, Edmonton, Alberta: Hurtig.

Goss, J. (1993) "Placing the market and marketing place: tourist advertising of the Hawaiian Islands, 1972–92," *Environment and Planning D: Society and Space*, 11, 6: 663–88.

Gowan, E. P. (1957) "A Quaker in buckskin," *Alberta Historical Review*, 5: 1–7, 24.

Graburn, N. H. H. (1995) "The past in the present in Japan: nostalgia and neo-traditionalism in contemporary Japanese domestic tourism," in R. W. Butler and D. Pearce (eds) *Change in Tourism: People, Places, Processes*, London: Routledge, 47–70.

Gregory, D. and Walford, R. (eds) (1989) *Horizons in Human Geography*, London: Macmillan.

Gregson, N. (1995) "And now it's all consumption?" *Progress in Human Geography*, 19, 1: 135–41.

Harding, A. (1970) Unpublished interview with Alice Harding by Susan B. Davies (May 18), Whyte Museum Oral History Program, Archives of the Whyte Museum of the Canadian Rockies, Banff, Alberta, Sl/52(A).

Hart, E. J. (1979) *Diamond Hitch: The Early Outfitters and Guides of Banff and Jasper*, Banff: Summerthought Publishing.

——(ed.) (1980) *A Hunter of Peace: Mary T. S. Schäffer's Old Indian Trails of the Canadian Rockies*, Banff, Alberta: Whyte Museum of the Canadian Rockies.

——(1983) *The Selling of Canada: The CPR and the Beginnings of Canadian Tourism*, Banff, Alberta: Altitude Publishing.

Henderson, K. A. (1991) *Dimensions of Choice: A Qualitative Approach to Recreation, Parks and Leisure Research*, State College PA: Venture Publishing.

Henderson, K. A., Bialeschki, M. D., Shaw, S. M. and Freysinger, V. J. (1989) *A Leisure of One's Own: A Feminist Perspective on Women's Leisure*, State College PA: Venture Publishing.

Hills, T. L. and Lundgren, J. (1977) "The impact of tourism in the Caribbean: a methodological study," *Annals of Tourism Research*, 4, 2: 248–66.

Hinman, C. (1915–1960) unpublished papers, Caroline Hinman Fonds, Archives of the Whyte Museum of the Canadian Rockies, Banff, Alberta, M236.

Holcomb, B. and Luongo, M. (1996) "Gay tourism in the United States," *Annals of Tourism Research*, 23, 3: 711–13.

Holloway, J. C. (1987) *The Business of Tourism*, London: Pitman.

Hughes, G. (1991) "Conceiving of tourism," *Area*, 23, 3: 263–7.

——(1992) "Tourism and the geographical imagination," *Leisure Studies*, 11, 1: 31–42.

——(1995a) "Authenticity in tourism," *Annals of Tourism Research*, 22, 4: 781–803.

——(1995b) "The cultural construction of sustainable tourism," *Tourism Management*, 16, 1: 49–59.

Jackson, E. L. and Burton, T. L. (eds) (1989) *Understanding Leisure and Recreation*, State College PA: Venture Publishing.

Johnson, R. (1986) "The story so far: and further transformations?" in D. Punter (ed.) *Introduction to Contemporary Cultural Studies*, London: Longman, 277–313.

Kinnaird, V. and Hall, D. (eds) (1994) *Tourism: A Gender Analysis*, Toronto, Ontario: Wiley.

Kipling, R. (1908) "Kipling in Collier," Mary Schäffer Warren Fonds, Whyte Museum of the Canadian Rockies, Banff, Alberta, M79.

Klein, J. (1993) "Blurring, cracking, and crossing: permeation and the fracturing of disciplines," in E. Messer-Davidow, D. R. Shumway and D. J. Sylvan (eds) *Knowledges: Historical and Critical Studies in Disciplinarity*, Charlottesville VA: University of Virginia Press, 185–211.

Kobayashi, A. and Mackenzie, S. (eds) (1989) *Remaking Human Geography*, London: Unwin Hyman.

Legault, S. (1996) "Caging Bear 16 is not a solution to the problem," *The Banff Crag and Canyon*, 31 July: 15.

Lehr, J. C. (1996) "One family's frontier: life history and the process of Ukrainian settlement in the Stuartburn district of southeastern Manitoba," *The Canadian Geographer*, 40, 2: 98–108.

MacBeth, R. G. (1924) *The Romance of the Canadian Pacific Railway*, Toronto, Ontario: Ryerson Press.

MacCannell, D. (1976) *The Tourist: A New Theory of the Leisure Class*, new edn 1989, London: Macmillan.

——(1989) "Introduction: special issue, semiotics of tourism," *Annals of Tourism Research*, 16, 1: 1–6.

McEwan, C. (1996) "Paradise or pandemonium? West African landscapes in the travel accounts of Victorian women," *Journal of Historical Geography*, 22, 1: 68–83.

McKee, B. and Klassen, G. (1983) *Trail of Iron: The CPR and the Birth of the West, 1880–1930*, Vancouver, British Columbia: The Glenbow-Alberta Institute/Douglas and McIntyre.

Mannell, R. C. and Iso-Ahola, S. E. (1987) "Psychological nature of the leisure and tourism experience," *Annals of Tourism Research*, 14, 3: 314–29.

Manning, E. W. and Dougherty, T. D. (1995) "Sustainable tourism: preserving the golden goose," *Cornell Hotel and Restaurant Administration Quarterly*, April: 29–42.

Mansfeld, Y. (1990) "Spatial patterns of international tourist flows: towards a theoretical framework," *Progress in Human Geography*, 14, 3: 372–90.

Marsh, J. (1982) "The evolution of recreation in Glacier National Park, British Columbia, 1880 to present," in G. Wall and J. S. Marsh (eds) *Recreational Land Use: Perspectives on its Evolution in Canada*, Ottawa, Ontario: Carleton University Press, 62–76.

——(1985) "The Rocky and Selkirk Mountains and the Swiss connection," *Annals of Tourism Research*, 12, 3: 417–33.

Mathieson, A. and Wall, G. (1982) *Tourism: Economic, Physical and Social Impacts*, London: Longman.

Matless, D. (1995) "Culture run riot? work in social and cultural geography, 1994," *Progress in Human Geography*, 19, 3: 395–403.

——(1996) "New material? work in cultural and social geography, 1995," *Progress in Human Geography*, 20, 3: 379–91.

Mieczkowski, Z. T. (1981) "Some notes on the geography of tourism: a comment," *The Canadian Geographer*, 25, 2: 186–91.

Mitchell, L. S. and Murphy, P. E. (1991) "Geography and tourism," *Annals of Tourism Research*, 18, 1: 57–70.

Moss, P. (1993) "Focus: feminism as method," *The Canadian Geographer*, 37, 1: 48–9.

Nast, H. J. (1994) "Opening remarks on women in the field," *The Professional Geographer*, 46, 1: 54–66.

Newell, W. H. (1990) "Interdisciplinary curriculum development," *Issues in Integrative Studies*, 8, 1: 69–86.

Nissani, M. (1995) "Fruits, salads, and smoothies: a working definition of interdisciplinarity," *Journal of Educational Thought*, 29, 2: 121–8.

Norris, J. and Wall, G. (1994) "Gender and tourism," in C. P. Cooper and A. Lockwood (eds) *Progress in Tourism, Recreation and Hospitality Management*, vol. 6, Toronto, Ontario: Wiley, 57–80.

Oakes, T. (1993) "The cultural space of modernity: ethnic tourism and place identity in China," *Environment and Planning D: Society and Space*, 11, 1: 47–66.

Palmer, E. (1996) "Seeing the sights: popular places to experience classic mountain scenery and adventure," *Where Rocky Mountains*, summer: 12–20.

Paris, I. (1984) unpublished interview with Herb and Ivy Paris by John Whyte, Whyte Museum Oral History Program, Archives of the Whyte Museum of the Canadian Rockies, Banff, Alberta, S1/149.

Pearce, D. G. (1979) "Towards a geography of tourism," *Annals of Tourism Research*, 6, 3: 245–72.

——(1989) *Tourist Development*, New York: Longman.

Przeclawski, K. (1993) "Tourism as the subject of interdisciplinary research," in D. G. Pearce and R. W. Butler (eds) *Tourism Research: Critiques and Challenges*, London: Routledge, 9–19.

Reed, K. A. L. (1856–1928) unpublished journals, Reed Family Papers, Hayter Reed Papers, Boxes 4–7, Archives of the McCord Museum, Montreal, Quebec.

Richer, L. K. (1995) "Gender and race: neglected variables in tourism research," in R. W. Butler and D. Pearce (eds) *Change in Tourism: People, Places, Processes*, London: Routledge, 71–91.

Rojek, C. (1990) "Baudrillard and leisure," *Leisure Studies*, 9, 1: 7–20.

Rose, D. (1993) "On feminism, method and methods in human geography: an idiosyncratic overview," *The Canadian Geographer*, 37, 1: 57–61.

Schäffer, M. (1907) "The valleys of the Saskatchewan with horse and camera," *Bulletin of the Geographical Society of Philadelphia*, 5: 36–42.

——(1908) "Among the sources of the Saskatchewan and Athabasca Rivers," *The Bulletin of the Geographical Society of Philadelphia*, 6: 16–30.

——(1910) *Untrodden Paths in the Canadian Rockies*, Minneapolis: Powers Mercantile Company.

——(1911) *Old Indian Trails: Incidents of Camp and Trail Life, Covering Two Years' Exploration Through the Rocky Mountains of Canada*, Toronto, Ontario: William Briggs.

Shaw, G. and Williams, A. M. (1994) *Critical Issues in Tourism: A Geographical Perspective*, Oxford: Blackwell.

Shields, R. (1990) *Places on the Margin: Alternative Geographies of Modernity*, London: Routledge.

Smith, C. (1989) *Off the Beaten Track: Women Mountaineers and Adventurers in Western Canada*, Jasper, Alberta: Coyote Books.

Smith, R. V. and Mitchell, L. S. (1990) "Geography and tourism: a review of selected literature, 1985–8," in C. P. Cooper and A. Lockwood (eds) *Progress in Tourism, Recreation and Hospitality Management*, vol. 2, Toronto, Ontario: Wiley, 50–66.

Smith, S. L. J. (1989) *Tourism Analysis: A Handbook*, New York: Wiley.

Squire, S. J. (1994a) "Accounting for cultural meanings: the interface between geography and tourism studies re-examined," *Progress in Human Geography*, 18, 1: 1–16.

——(1994b) "The cultural values of literary tourism," *Annals of Tourism Research*, 21, 1: 103–120.

——(1994c) "Gender and tourist experiences: assessing women's shared meanings for Beatrix Potter," *Leisure Studies*, 13, 4: 195–209.

——(1995a) "In the steps of 'genteel ladies': women tourists in the Canadian Rockies 1885–1939," *The Canadian Geographer*, 39, 1: 2–15.

——(1995b) "Travels in interdisciplinarity: exploring integrative cultures, contexts and change," *Issues in Integrative Studies*, 13, 1: 79–99.

Stember, M. (1991) "Advancing the social sciences through the interdisciplinary enterprise," *The Social Science Journal*, 28, 1: 1–14.

Strauss, A. L. (1987) *Qualitative Analysis for Social Scientists*, New York: Cambridge University Press.

Swain, M. B. (1995) "Gender in tourism," *Annals of Tourism Research*, 22, 2: 247–66.

Swoboda, W. (1979) "Disciplines and interdisciplinarity: a historical perspective," in J. J. Kockelmans (ed.) *Interdisciplinarity in Higher Education*, University Park PA: Pennsylvania State University Press, 49–91.

Timothy, D. and Butler, R. W. (1995) "Cross-border shopping: a North American perspective," *Annals of Tourism Research*, 22, 1: 16–34.

Tocher, M. (1977) unpublished interview with May Tocher by Carson Wade and Lisa Casselman (September 23), Parks Canada/Yoho National Park Sound Recording Project, Archives of the Whyte Museum of the Canadian Rockies, Banff, Alberta, NT144–1, Accn. 3381.

Towner, J. (1984) "The Grand Tour: sources and a methodology for an historical study of tourism," *Tourism Management*, 5, 4: 215–30.

——(1988) "Approaches to tourism history," *Annals of Tourism Research*, 15, 1: 47–62.

——(1994) "Tourism history: past, present and future," in A. V. Seaton (ed.) *Tourism: The State of the Art*, Toronto, Ontario: Wiley, 721–8.

Tunbridge, J. E. and Ashworth, G. J. (1996) *Dissonant Heritage*, London: Belhaven.

Unwin, E. (1970) unpublished letter from Ethel Unwin to Maryalice Stewart (June 18), Archives of the Whyte Museum of the Canadian Rockies, Banff, Alberta, M178.

Urry, J. (1990) *The Tourist Gaze: Leisure and Travel in Contemporary Societies*, London: Sage.

——(1995) *Consuming Places*, London: Routledge.

Veijola, S. and Jokinen, E. (1994) "The body in tourism," *Theory, Culture and Society*, 11, 3: 125–51.

Waitt, G. and McGuirk, P. M. (1996) "Marking time: tourism and heritage representation at Millers Point, Sydney," *Australian Geographer*, 27, 1: 11–29.

Wall, G. (1996) "Sustainability in tourism and leisure," *Annals of Tourism Research*, 23, 1: 224–5.

Warren, S. (1993) "This heaven gives me migraines: the problems and promise of landscapes of leisure," in J. Duncan and D. Ley (eds) *Place/Culture/Representation*, New York: Routledge, 173–86.

Wolfe, R. I. (1951) "Summer cottagers in Ontario," *Economic Geography*, 17, 1: 10–32.

6

TOURISM AND THE CONSTRUCTION OF PLACE IN CANADA'S EASTERN ARCTIC[1]

Simon Milne, Jacqueline Grekin and Susan Woodley

Introduction

Since the early 1980s several Inuit communities in Canada's Baffin region have turned to nature- and culture-based tourism as a source of much needed income and employment (Anderson 1991; Hamley 1991; Hinch and Swinnerton 1993). Driven in large part by the continued marketing efforts of the Government of the Northwest Territories (Hamburg and Monteith 1988; GNWT 1991, 1993), Tourism Canada (ISTC 1990), a range of southern tour operators and the communities themselves, visitor numbers have more than doubled since the early 1980s, reaching approximately 3,000 by the mid-1990s (GNWT 1992; Milne *et al.* 1997).

In this chapter we examine how tourism spaces in this region of Canada's eastern Arctic have been constructed. We begin by providing a brief review of some of the key themes to have emerged in the growing literature on the transformation of landscapes into tourist places by promotional agencies, the tourism industry and visitors themselves. We also stress that if we are to understand the construction of tourist landscapes in the Baffin region we must look beyond these "outside" actors, and also pay attention to the evolving ability of "insiders" (local communities and residents) to shape tourism spaces.

Our empirical analysis begins with an overview of the various external forces that have helped to shape tourist perceptions of the region. We examine some of the "myths" that have helped to shape Canadians' broad conception of the Arctic, focusing in particular on the role of film, radio and literature. We then move on to examine how these images have been perpetuated (and in some cases subverted) by the packaging and promotion conducted by Southern tour operators, and various levels of government.

A review of tourist perceptions of, and reactions to, their travel experiences (drawn from two visitor surveys conducted in 1992 and 1993) reveals, in very broad terms, the degree to which the "idea of north" has been internalized by travelers, and how this influences some of the impacts associated with tourism development in the region. In particular, our focus is on the idealized view that tourists often hold of the Inuit and their relationship with the surrounding landscape, a view that stands in stark contrast to realities of everyday life in the region.

Consequently, we also examine the degree to which communities can mold the creation of tourist landscapes that better mesh with the realities of northern life. Resident surveys reveal some of the areas of actual and potential conflict that exist between tourist and resident perceptions of the Arctic landscape—especially in relation to the consumption of wildlife resources. However, new political structures (Nunavut) as well as community controlled activities (brochures, visitor programs) and new technologies (the Internet) offer the potential for local communities to play a significant role in the construction of a tourist landscape, and to aid in the development of an industry that is more appropriate to the needs, aspirations and way of life of the Inuit people.

The construction of tourist places, "once a certain idea of landscape, a myth, a vision, establishes itself in an actual place, it has a peculiar way of muddling categories, of making metaphors more real than their referents, of becoming, in fact, part of the scenery" (Schama 1996: 61). Tourism is not just a commercial enterprise, but an "ideological framing of history, nature and tradition, a framing that has the power to reshape culture and nature to its own ends" (MacCannell 1992: 11). The role that tourism can play in transforming collective and individual values is inherent in the idea of commoditization (Britton 1991; MacCannell 1989: 21). Cultural landscapes are viewed and shaped as commodities—commodities that can then be consumed by potentially malleable consumers.

Consumers are tutored to see a particular representation of a commoditized landscape in a number of ways, and via a number of actors (Dilley 1986; Cohen 1993; Dann 1996). Initial tourist preconceptions and expectations of a destination are shaped by "maps of meaning" (Jackson 1989) that are often created by broader forces exogenous to the tourism industry (e.g. film, literature the educational system, etc. See also Urry 1990). The tourist industry also creates images of place via advertising, and these in turn create expectations on the part of the visitor (Hall 1994: 178).

Finally the travel experience itself generates memories that tourists take back home with them, and which will influence not only their own perceptions of place, but also those of friends, relatives and colleagues. Thus slide shows, photos and word of mouth represent another set of influences that can shape perceptions of the tourist landscape. As Squire (1994a: 7) notes, when "cultural texts are read, meanings are also incorporated into lived cultures or everyday life, eventually contributing to new moments of production, new textual forms, and ultimately, new readings."

Much of the literature on the construction of tourist landscapes in peripheral regions and areas where indigenous peoples live, has focused on the fact that this commodification process involves the elaborate creation of "fantasy." An unreal world is created that will meet the tourist's need to partake in "imaginative hedonism" and pleasurable experiences (Campbell 1995: 119), providing an opportunity to get away from the sites of everyday routine and into the extraordinary (Urry 1990, 1995). Tourist advertisements thus remake places in a special way, providing a representation of reality that exists in opposition to the everyday life of most participants in mass, urban, industrial societies (Hummon 1988: 200).

Of particular relevance to this chapter is tourism's use of nature as an oppositional force to everyday urban life. As nature and society have become increasingly separated, tourism has come to represent an important mechanism through which to define what is natural (Urry 1990: 98; Lash and Urry 1994; Wilson 1992). Indeed Bell (1996: 40) argues that a romanticized appreciation of nature is most likely to flourish in cultures with highly developed technologies.

This need to create the "other" is argued to distort reality rather than providing a depiction of the true social and economic situation in the destination (Britton 1979; Silver 1993). It also brings with it important implications for the development process and the creation of the materiality and social meaning of places (Britton 1991). Thus Britton (1982) argues that the advertising strategies of metropolitan corporations will lead visitors to seek the types of experiences and standard international facilities associated with mass, globalized tourism. Small-scale, locally owned enterprises are either relegated to activities which lie beyond the immediate interests of larger companies, or find roles as subcontractors. The outcome of this process is tourist industries that satisfy the commercial imperatives of overseas interests and rarely meet local development needs. Hall (1994: 178), drawing on Urry's (1990) notion of the tourist gaze, sums up this vicious cycle of commodification as follows:

> Tourism redefines social realities. Advertising creates images of place which then create expectations on the part of the visitor, which in turn may lead the destination to adapt to such expectations. Destinations may therefore become caught in a tourist gaze from which they cannot readily escape unless they are willing to abandon their status as a destination.
>
> (Hall 1994: 178)

While this literature has much to recommend it, and has clearly helped us to forge some important links between tourism research, the understanding of the commodification process and the evolving field of cultural studies, it is important to note that a number of key gaps remain to be filled: gaps that will, to some extent, provide the focus for the development of this chapter.

First, there is clearly a need for more empirical research to be conducted in the cultural studies arena, particularly in the areas of overlap with tourism studies (Jackson 1989; Squire 1994a: 8). In particular we need to explore tourist, resident, government and private sector visions of tourism spaces, and how these landscapes are commodified and consumed. Much of the recent work in this area has focused on the producers of the tourist experience rather than on its consumers. Little research has, for example, been conducted on the cultural meanings and values that visitors associate with their activities. At the same time researchers have typically mediated and interpreted how visitor experiences are constructed and communicated, with the voices of ordinary visitors rarely being heard (Squire 1994b: 107).

We would argue that it is equally important to better understand community and resident perceptions of tourist space, and the degree to which this conflicts with indigenous/local conceptions of landscape. In particular we must attempt to gain a better understanding of how communities can subvert and shape the tourism industry's portrayal and use of the physical and cultural landscape to better meet local needs (e.g. Preston 1996). It is only with this type of understanding that we can move beyond the vicious cycle described above and reach a resolution to the apparent incompatibility between the commodification conducted by the tourism industry and the needs and wishes of local peoples (see Hollinshead 1996: 310).

Finally, the construction of tourism must be considered within the context of broader changes that are influencing the economy and popular culture (see also Urry 1990, 1994). There is, no doubt, for example, that shifts in the structure and organisation of the tourism industry, (e.g. the greening of the industry) will not only alter the nature of tourist services, but will also change the relationship between producers and consumers of tourism products, and "how the meanings of the tourist experience are negotiated by various agencies" (Squire 1994a: 8). A great deal of work remains to be done here, especially in the important area of grafting current theories of economic change onto our understanding of the tourism industry and the production of tourist places (see Milne forthcoming). We must also look carefully at how the development of new technologies and modes of global communication, especially the Internet, are influencing traditional methods of developing and disseminating tourist images. These new distribution technologies may, in fact, allow communities and local peoples to take a more proactive role in the creation of tourist places.

Creating the "idea of north"

The Arctic of outsiders is a landscape of the mind, shaped more in the imagination by reading than by experience and perception . . . for those who have never been to the Arctic, this is the only northern reality they know.

(Moss 1994: 23)

Media has played an important role in shaping people's perceptions of northern landscapes. Any attempt to review this role must of course be selective, and in this section we focus on those films, books and radio shows that have played a particularly important role in influencing Canadian perceptions of the north. Our choices are also based, to a limited extent, on some of the background reading and viewing recommendations provided by the tour operators that bring visitors to the Baffin region.

Nanook of the North (Flaherty 1922) represents one of the earliest and most famous attempts to capture Inuit life on film. It presents an enduring and romantic image of a primitive and happy people that can survive in the face of the utmost adversity:

The sterility of the soil and the rigor of the climate mean that no other race could survive, yet here, utterly dependent upon animal life, which is their sole source of food, live the most cheerful people in all the world—the fearless, lovable, happy-go-lucky Eskimo.

(Flaherty 1922: 33)

This is a myth that has continued to pervade the public image of the north for much of this century and which still persists, to some extent, to this day (Brody 1991).

The image of the Inuit as "noble savage" is also stressed through Nanook's abilities as a formidable hunter. The collective family unit is seen as central to Inuit survival, and relationships between its members are portrayed as harmonious. This final point is driven home at the film's end, with the camera switching back and forth between the inside of the igloo (with Nanook and his family warm and asleep) and the bitter environment without.

Between Two Worlds (Greenwald 1990) focuses on a dominant image in much of the post-1960s media coverage of the Arctic, that of a place whose people and resources have been exploited and conquered by the South (Brody 1991: 90–1). The film details the "rise and fall" of Joseph Idlout, whose fame started when he and his camp were featured on the back of Canada's two-dollar note. He went on to be featured in several newsreels and books as a great hunter, fighting cheerfully against enormous environmental odds. *Between Two Worlds* details Idlout's transformation into a bullying drunk, with his shattered self-respect providing a metaphor for broader change in Inuit culture, especially the present-day problems of alcohol and solvent abuse in northern communities. The film ends with a poignant shot of the past—Idlout and his son Paniloo happy and laughing around their igloo—the grown son then describes a song they used to sing about going back to the way it used to be; but he knows they can never go back.

The big budget Hollywood film *Shadow of the Wolf* (Dorfman 1993) again emphasizes the conflict between two very different cultures. The lead character, Agaguk, is an angry man—angry at the intrusion of white law and culture. The Inuit characters are often shown as drunk or violent, and are presented as being easily manipulated by whites. While family and community life are not presented through rose-tinted spectacles, the strains created by everyday life in the Arctic are nevertheless often conveyed via the metaphor of fighting dogs, rather than through conflicts among the Inuit themselves.

The exploitation of the north is addressed in somewhat broader terms in "The Idea of North," Glenn Gould's (CBC 1992) radio documentary (since released on CD). This "soundscape" traces the impact that the north has had on five characters who have spent time in the area, and is an attempt to move beyond the romanticized image of the north that was presented to Canadians during the post-World War II years. The documentary unfolds on a train traveling between southern Ontario and Churchill, Manitoba. Towards the end of the piece, the characters describe how they envision the north of the future. One thinks it will look like a southern suburbia, with roads and malls; another theory is that it will become a vast playground for outdoor recreation. In conclusion it is stressed that the north can be seen as "the moral equivalent of war": the battle, however, is no longer with Mother Nature, but human (southern) nature, which is gradually creeping north.

These themes of exploitation are also reinforced, to some extent, in two of the most popular literary interpretations of the north to have emerged in the recent decades: Farley Mowat's "Canada's North Now: The Great Betrayal," and Barry Lopez's "Arctic Dreams." Both writers attempt to adopt almost an "insider" perspective in their respective volumes. Mowat (1976) stresses that:

> This north, this Arctic of the mind, this frigid concept of a flat and formless void of ice and snow congealed beneath the impenetrable blackness of the polar night, is myth! Behind it lies a real world, obscured in drifts of literary drivel and buried under an ice weight of obsessive misconceptions; yet the magnificent reality behind the myth has been consistently rejected by most Canadians since the day of our national birth.
>
> (Mowat 1976: 9)

He goes on to argue that a new conception of the north has come into being with southern Canadians being hoodwinked into seeing the north as a source of riches in an otherwise useless wilderness, resources which ought to be immediately drained off to the south (1976: 12). Meanwhile, Lopez (1986) examines the influence that the Arctic landscape exerts over the human imagination, and how our desire to use the natural resources of the north has shaped our conception of the Arctic. He places an emphasis on the knowledge of indigenous peoples and presents an overview of the Arctic landscape which is more human-oriented than much of the previous literature dealing with the north.

While both Lopez and Mowat try to be more realistic and sensitive in their interpretation of the Inuit and their changing way of life, the problem with this literature is that the authors attempt to downplay their role as outsiders.

One of the few writers to have effectively addressed this issue is the poet Al Purdy (1967) whose words admit, with humility and pride, the alien perspective of his exposure to the Arctic and, implicitly, the distance that lies between his own realities and the world he writes about (see Moss 1994). In his poem *Dead Seal* (1967: 58–9) this outsider perspective is revealed through the metaphor of a boat ride with an Inuit hunter and a freshly killed seal. Purdy describes how intrigued he is by the animal and how much he wants to touch it. He then decides to leave it alone and that he does not need to prove anything by touching it, feeling the experience would be disgusting and disturbing. But then he changes his mind and wonders what has overcome him.

More recent attempts to come to terms with the north have attempted to reflect indigenous perceptions of the landscape (Moss 1994: 23), and the difficulties and dangers inherent in attempting to create some degree of understanding among outsiders. In particular we are beginning to see attempts to reinterpret the "idea of north" through the role of indigenous knowledge and local conceptions of "wilderness." As Kaltenborn (1995: 107–8) argues: "A wilderness concept and framework for the polar regions must be heavily influenced by the indigenous peoples who live there, and it must also be articulated in a language that is understood in industrialized societies."

The creation of the Arctic as a tourist commodity

In an attempt to gain some sense of the way in which tour operators to Baffin Island package, promote and develop their products, we now draw on a survey of eleven companies based in southern Canada and the United States that was conducted during 1992–3 (see Grekin 1994; Grekin and Milne 1994). Two inbound operators based in Iqaluit (the regional gateway to the Baffin region) were also interviewed in July 1992. This material is further supplemented by a more recent review of operator brochures.

A content analysis of brochures and related promotional materials reveals that physical landscape, wilderness and environment receive most attention (icebergs, fiords, tundra), wildlife comes in a fairly distant second (polar bears, whales, musk-ox) and the Inuit and their culture are ranked third. There is a large number of adventure operators who place virtually no emphasis on people, culture and community. A smaller number attempt to build a community/culture dimension into some aspect of their product, often in the form of a brief community stopover just prior to departure. Where culture is mentioned in these cases, very little attention is given to the idea of any form of interaction, with most brochures focusing on "viewing" and "exploring" the communities concerned.

The bulk of the material dealing with local people and culture comes from operators who deal specifically with community-based tourism, and have little direct nature or adventure material in their package. Themes developed in these products at the community level include Inuit art, archaeological sites and the history of "outside" influences on the area (explorers, whalers, fur traders, missionaries). The lifestyles and customs of the Inuit are featured less often, but are in evidence on occasion, with a focus on dog-sledding, building igloos, kayaking and, to a lesser degree, the consumption of country foods.

While most tour operators provide their clients with some pre-trip information about the nature and culture of the region, it is generally rather limited in scope. Seven of the operators surveyed in 1992 claimed to provide detailed information on these subjects, with some of these companies preparing reading lists (often including Lopez and Mowat). In other cases standard government pamphlets are sent to prospective visitors. Three of the operators indicated that their clients were professional people who could educate themselves without the assistance of the operator.

Interestingly, only one of the operators surveyed prepared tourists explicitly for the possibility that they may see seal or narwhal hunting and explained the importance of this activity for Inuit. Another tour operator took the opposite approach, by taking pains to have his guides avoid all evidence of hunting activity. At the same time the important implications of the new Nunavut Territory (see below) are rarely, if ever, mentioned in brochures. In simple terms, most tour operators do not appear to be taking a proactive approach to educating their clients about the realities of Arctic life. The tourist landscape they present remains a very idealized one—although it does not reach the level of fantasy creation alluded to in much of the literature (see Hollinshead 1992).

Government

Since 1983 the style and pace of tourism development throughout the Northwest Territories has been guided by a formalized "community based" strategy which emphasizes an industry that is environmentally and culturally sustainable, broadly distributed between communities, and which yields maximum possible economic benefits for residents, particularly those of small- and medium-sized communities (GNWT 1983; Hamburg and Monteith 1988).

Our review of government promotional materials reveals a greater emphasis on the human and community dynamics of the north than is found in the tour operator material. Most government publications emphasize that the north is an area of cultural and societal change, where people live in an interdependent fashion with the natural resource base that surrounds them. They also tend to emphasize a tourism that involves more in the way of community visits and interaction with local economic structures. This clearly represents an attempt to spread the economic benefits of the industry.

The *Baffin Handbook* (Hamilton 1993), production of which was subsidized by government funds, and which is largely sold through the local Tourism Association offices, undoubtedly provides the most comprehensive and realistic perspective on the Baffin region's tourist product. The book's overall approach is illustrated in its discussion of the pace of life: "remember that life moves at a slower pace in the north; don't expect everything to move quickly or on a precise schedule" (17–18), the high cost of outfitting services, provisions and accommodation: "high prices for modest accommodation" (20), the "uninspired" style of cooking available in most hotels, and the difficulty visitors may have in finding and sampling "country" foods. The book also provides behavioral guidelines for visitors, who are requested not to photograph Inuit while eating, not to ridicule their eating habits and not to think of them as "historic pieces" that are always smiling, living in snow houses and are somehow "frozen in time."

Nevertheless some government materials do continue to perpetuate the myths of the north—particularly the "unchanging" nature of communities. Thus the Keewatin region brochure, *Canada's Arctic, the Last Frontier*, states: "Your visit to Whale Cove is a visit to a charming traditional community of just 210 souls. Here the bond to the land is still strong. This may be your last chance to see yesterday's disappearing Arctic." Meanwhile the Baffin Tourism Association states in one of its promotional documents: "Let us show you a place still untouched by man. Some say it conjures up the exotic, the unexplored and a sense of timelessness."

Tourist perceptions

The degree to which tourist perceptions of the eastern Arctic as a tourism landscape have been influenced by media, operator and government promotional materials is of course hard to ascertain. While we cannot easily evaluate which types of media and information have helped to form travelers' basic ideas of the north's physical and cultural landscapes, we can look at what factors are significant in shaping their decision to visit the region and to consume certain elements of the tourist product.

Our tourist surveys of 1992–3 (departure survey of 344 respondents conducted at Iqaluit airport; see Grekin and Milne 1996; Milne *et al.* 1997) reveal that word of mouth is, overall, the single most important source of information influencing the holiday decision making process for those traveling independently (Milne *et al.* 1997). This grouping also relies heavily on government promotional materials. For those on a package, tour operator information is the most important factor, with word of mouth coming in a close second. It is interesting to note that over three-quarters of the tourists surveyed also stated that they would recommend Baffin Island to a friend. This reinforces our need to better understand the role of tourists in reproducing and reshaping tourist landscapes.

We also asked tourists to rank their motives for traveling to the region. The category "environment/scenery" was considered the most important attraction by both package and independent tourists. "Wilderness" was rated second among independent tourists, with "Inuit people/cultural experience" also featuring significantly. "Inuit people/cultural experience" was the second choice for package tourists, with "wildlife" and "wilderness" also rating strongly. Other surveys of tourist motivations in traveling to the Baffin region tend to confirm these findings (Keller 1982: 103; Acres International Ltd 1988).

Respondents were also asked to list the features of their trip which they found satisfying/least satisfying. While tourists are highly satisfied with the region's natural attractions and the "friendliness of people," they were far less satisfied with the opportunity provided for real cultural and social interaction (see Grekin and Milne 1996; Milne *et al.* 1997). An earlier study of tourism development in Pangnirtung also revealed a high level of curiosity about the social, cultural, economic and political issues facing Inuit today, and a lack of satisfaction with the opportunity to sample country foods, meet carvers or to have any contact with residents (Reimer and Dialla 1992).

Some of the written and oral comments provided by our respondents reveal this desire to gain more access to Inuit culture—while a few wanted to see "staged" cultural performances and ceremonies, several visitors also expressed a keen interest in viewing "real life" activities. Thus one independent traveler:

> would have liked to hear some Inuit throat singing, to learn more about how they hunt, and use the seal products, etc., but the opportunities never arose. I would have loved to meet the elders, and hear their stories, but again, short of barging into their meetings, how do you meet them? . . . The educational programs aren't in place. The ones that are, are too expensive.

One group of three tourists who visited Pond Inlet in 1993 made specific mention of their unfulfilled desire to view hunting and identified a possible explanation:

> We quickly realized that tourists were regarded as Greenpeace members. We asked several times to join a seal hunt but noticed lots

of reluctance and got negative answers. . . . I regret this fact. It would have been very interesting for us.

It is interesting to note that those who were offered exposure to Inuit culture usually thought highly of the experience.

Without reservation, the highlight of the trip was the opportunity to observe (and participate in) the life of the Inuit "on the land." Although the observation of wildlife at the floe edge was the intended purpose, this took a secondary place to the "Inuit life" factor once we were on the ice with our Inuit guides.

While there is clearly a great deal of interest among visitors in Inuit culture, MacCannell (1992: 26) posits that this type of interest in "primitive" cultures holds largely negative connotations. He argues that the postmodern tourist's gaze is highly critical and value-laden, focusing on attempts to capture the essence of "unspoiled nature and savagery." This results in a tendency for tourists to impose on cultural groups their own view of what constitutes a "traditional" ethnic identity. Thus ethnic tourism is considered to be "the mirror image of racism" (*ibid.*: 170). In a similar vein, Wenzel (1985, 1991) has similarly noted that within the animal rights movement, there is "a resentment that Inuit have not risen above the crass consumerism of southern society."

There is evidence that some tourists do indeed find some elements of Inuit culture objectionable, particularly modernized hunting activities (skidoos, rifles) and the "wastage" of animal remains. For example, in a letter to the editor of a nature magazine, a tourist to Pond Inlet stated: "I am left with images of waste . . . by men with rifles and snowmobiles" (Nature Canada 1988). Several of the tourists surveyed provided similar impressions, although they were by no means in the majority. A German student, on a package tour to Pond Inlet, objected to the hunting of narwhal, which he felt was excessive. A group of Americans camping and kayaking near Pangnirtung noted that "the shooting of seals is loud and obtrusive." The issue of waste was also noted by a Canadian visitor who was "shocked by the apparent total disregard of the northerners for the environment and for conservation of the natural resources (polar bear, narwhal, seal, etc.)."

To develop a clearer understanding of tourist attitudes toward hunting, the 1993 airport survey asked whether respondents had witnessed any such activity, and what they felt about the experience. Nearly one third of respondents chose not to complete this question, leaving ninety-nine usable replies. For those who did view hunting, just one individual described it as a negative experience, with the other twenty describing it as positive. Over 40 percent of the group who saw no hunting (seventy-eight) claimed that they would have liked to have had this experience, but that the opportunity was lacking.

Community and the construction of place

Various studies have shown that local Inuit generally have a positive attitude toward the tourism industry, and there is relatively little evidence to show that communities have found it to be a disruptive force in either a cultural or environmental sense (Reimer and Dialla 1992; Grekin and Milne 1996). These relatively positive perceptions are closely tied to a belief that the industry can generate seasonal income and jobs. In most cases these perceptions have proven to be well-founded, with the industry playing an important role in bolstering the economies of those communities that have made the greatest inroads in attracting visitors, such as Pangnirtung or Pond Inlet (Reimer 1989; Grekin 1994). Even in communities where the economic performance appears to have been less impressive, such as Clyde River (Nickels *et al.* 1991) and Cape Dorset (Milne *et al.* 1995) community support is relatively strong, perhaps because of the potential benefits that are seen to eventually accrue from the industry.

Community surveys have, however, revealed that some Inuit are concerned that tourists are going straight into the "wilderness" and do not interact much with locals (Grekin and Milne 1996; Milne *et al.* 1995). As a result there is concern that visitors may misunderstand the Inuit way of life and its intimate links with the surrounding resource base, and then go on to spread "false rumours," especially about the use of wildlife, in the south. For this reason a large number of community members in the various communities surveyed want tourists to learn about Inuit culture, and would like tourists to meet with elders and local people in general to get a "real" perspective on the Arctic's physical/cultural landscape.

If they are to overcome these problems it is clear that the Inuit must be able to develop their own images of culture and history—to commodify the Arctic landscape in such a way that it helps create a tourism industry that meets local needs and expectations (see Hollinshead 1992: 56). We now move on to provide some brief examples of the efforts being made to achieve local input into the formation of the Baffin tourist landscape. We show that several important factors are playing a role in attempts to achieve this goal, and that they range in scale from the global (the Internet) to the regional (land claims agreements) and finally to the local (community initiatives).

The Internet

Kulchyski (1989: 54) argues that the potential for Inuit to define themselves in an increasingly commodified society "will depend on their ability to subvert capitalist economy, technology, images, and institutions." In this respect the continued growth and development of the Internet may provide the Inuit, and peoples/localities everywhere, with one of their best chances to reassert difference in an increasingly globalized setting (Friedman 1996; see also Bell 1996: 21).

In November 1994 a symposium called "Connecting the North" was organized to discuss the best options available for communities to link into this growing communication medium, and since this time the presence of Canada's Arctic on the "Net" has grown significantly (Friedman 1996). While many of these sites exist predominantly with the aim of inter-community contact, the details on Inuit life and local landscape which are provided represent valuable information through which to directly (and indirectly) shape tourist perceptions of the north.

The key tourism-related site is the NWT Virtual Explorers' Guide (http://www.edt.gov.nt.ca/guide/index.html) which offers a variety of information and tour options. This site offers access to information on the history of the land, vacation planning, different types of tours, parks, and also contacts for Western and Eastern tourism operators.

Community visits are widely encouraged: "traditional communities, steeped in history and Inuit culture . . . many Inuit tour operators [are] willing to take you out to see their land and wildlife." In fact the Guide provides an entire history of communities in the eastern Arctic from the 1950s to the present. The site also tells the potential tourist how to act and what to expect: "You will find our people friendly, but sometimes shy. Ask permission before you take pictures. . . . Be careful not to be intrusive, and you'll find people quite willing to talk." One marked difference between standard tour operator and government brochures and this Internet material, is that the latter seems to place far more emphasis on "our land," and stresses pride in way of life:

> We invite you to visit our land and share in our pride of culture and our love of the land. In the eastern NWT, spend time with an Inuit elder as he frees the image of a polar bear from soapstone. Imagine the pull of a giant Arctic char on your line or gaze in awe at the great herds of caribou as you paddle a remote river.

Another common theme is the blend of modernity with tradition: "It is a land where the people are still very much in touch with their past, but also caught up in the excitement of the birth of their own land, their beloved Nunavut" (NWT Virtual Explorers' Guide). Thus the Web appears to offer considerable potential for the Inuit of Baffin island to provide their own interpretation of the Arctic landscape to potential tourists. The highly educated, high-income profile of most visitors to the region (see Milne *et al.* 1997) also makes it extremely likely that these are exactly the types of consumers who will have access to the Internet (see Hull 1996; Milne and Gill forthcoming).

We should not pretend, however, that the Internet provides anything like a total solution. Cost and technological barriers persist as important impediments to peripheral regions that are attempting to reach and influence the market place through this new distribution technology (see Milne 1996; Hull 1996). At the same time it is important that communities and locals have direct input into the website development process, otherwise they run the risk of again placing the construction of "place" in the hands of outsiders.

Nunavut

On 25 May 1993 the Government of Canada and the Inuit of the Nunavut Settlement Area signed the Nunavut Land Claims Agreement. From 1 April 1999 a provincial/territorial government will speak on behalf of a large group of native people for the first time (Pelly 1993). Although the Agreement makes no explicit reference to tourism, provisions relating to land-use planning, park and protected areas establishment, and structures providing Inuit with greater control over the development of their resources, will have a direct bearing on the industry (see Woodley 1996).

The Nunavut government views tourism as being among its top priorities in terms of community economic development. A series of new administrative arrangements, including the amalgamation of three existing tourist regions (Baffin, Kitikmeot and Keewatin), will create one tourism entity which will fall under the umbrella of Nunavut Tourism (Woodley 1996). While the basic objectives of Nunavut Tourism closely resemble those of the GNWT's (1983) "Community-Based Tourism" strategy, the main difference is that the coordinating body will be controlled and operated by Inuit. In real terms this may mean that Inuit will have greater control over the regional development and commodification of the tourism resource. It is still too early, however, to see how meaningful the transition will really be. There is clearly much research to be done on the degree to which land-claims processes, and the ability of indigenous peoples to gain more control over their resource base and political destinies, will translate into the creation of new, more locally appropriate forms of tourist space.

Community initiatives

Irrespective of the influence of Nunavut and the Internet, it is clear that communities still have to develop their own presence in the market place, and play a direct role in shaping and molding the tourist spaces around them. In this respect many communities use brochures as marketing tools, which can enable them to both attract tourists and educate them about the landscape they are about to encounter. While some brochures are quite basic, emphasizing practical information on location of facilities, and basic historical, physical and recreational information, others present an array of information on the history, economy and culture of the specific locality.

A good example of the ability of communities to define and mold tourist expectations is to be found in the 1994 brochure developed by the Hamlet of Grise Fiord. The project was initiated, defined and controlled by the Hamlet Council and the local Inuit cooperative. While research and writing was contracted to a non-resident, they were given strict instructions on what to include. The brochure focuses on a number of themes, including community history, the history of the Royal Canadian Mounted Police in the region, local flora and fauna, and general community information. The brochure also establishes codes of conduct which visitors are requested to observe. A note requesting that tourists do not walk on nearby archaeological features, or remove any artifacts, is included. An explanation of the seemingly harsh treatment of dogs is also provided, alongside a great deal of emphasis on wildlife as both spectacle and resource. The tourist's attention is drawn to the fact that ringed seal, narwhal, beluga and musk-ox are vital elements of the local diet. Conscious of the potential problems associated with tourist perceptions of animal "waste," the brochure goes on to state:

> Evidence of recent hunting expeditions scattered about the beach testify to the continued importance of hunting in Grise Fiord. It is also an indication of the different perception and use of space in northern communities. There are no white picket fences here to delineate personal property. But do not be deceived: those muskox horns lying on the beach have not been abandoned; that seal is not going to waste: these items belong to someone and will be claimed when needed. You are trusted to respect this fact.
>
> (Grise Fiord 1994)

The key informants in the development of the brochure were the mayor, who is also an outfitter and guide, the Renewable Resources Officer, the president of the Hunters and Trappers Association (HTA) and the manager of the Cooperative. Two elders also provided information about community history. This fact does raise the important issue that any local initiative will itself tend to reflect local power structures, and that the images created may not necessarily reflect the needs and wishes of all community members. Thus in the case of Grise Fiord the voices of young people and women were not really heard in the construction of the brochure. This points to the ongoing need within tourism research to better understand "community" and the power structures that characterize its formation (Milne forthcoming).

Brochures are, of course, only one way in which communities can attempt to shape tourist perceptions of local landscapes. Some communities are developing "meet and greet" programs, with residents taking recently arrived visitors on guided tours of the community—to point out the "do's and don't's" of being a tourst in a northern setting, and explaining details of the Inuit way of life. Other communities have made use of Federal and Territorial government funds to develop community cultural centres—which serve the primary purpose of providing community members with a central meeting place, while also providing opportunities for visitors to view community history, and in the case of Pangnirtung, to meet with and listen to elders.

Conclusion

The current boom in cultural tourism is taking advantage of tourists' thirst for knowledge about aboriginal peoples' traditions and present lifestyles, and their own quest for cultural souvenirs (see Greenwood 1989; Smith 1994). However, great care will have to be taken to ensure that the tourism product is appropriate to indigenous values, that it reflects local community wishes, and that aboriginal people themselves obtain the economic benefits that they desire. This may well be the major challenge facing Arctic tourism in the next decade (Johnston 1995). It is also clear that the construction of tourist place through the broader media, tour operator brochures and government marketing initiatives, plays a significant role in determining tourists' understanding and consumption of the landscapes they visit, and will influence the ability of communities to meet this challenge.

We have shown in this chapter that the "idea of north" has been constructed, to some extent, by a range of external agents, and that tourist perceptions of the eastern Arctic tend to reflect this process of commodification. We have also argued that tourists themselves must be seen as important actors in the perpetuation and recreation of tourist landscapes through their collecting, reading and communication of tourist images. The chapter has revealed that the gap between tourist and local perceptions of the landscape and its consumption can create problems for local communities.

A focus on wilderness and scenery at the expense of community and culture can, for example, lead to limited economic linkage-creation and also a lack of understanding of local cultural traits and way of life. We have shown in particular that Inuit use of wildlife resources is often misunderstood by tourists, causing fears on the part of community members that negative images will be relayed to the south—exacerbating past misunderstandings created by the animal rights movement (see Wenzel 1985; 1991). At the same time we have pointed to the fact that tourists are generally unsatisfied with their opportunities to learn more about Inuit culture and way of life, and that community members themselves want the opportunity to present their perspective of the Arctic landscape.

If the Inuit are to escape the "suffocating straight-jacket of enslaving external conceptions" (Hollinshead 1992: 44) that is argued to characterize indigenous tourism in other parts of North America, it is vital that they play a role in interpreting history and landscape from their own perspective. We have shown that global changes in communication technologies (the Internet), the creation of greater political autonomy through land-claim agreements (Nunavut) and individual community initiatives can all play an important role in assisting communities to meet this goal.

At the same time, however, ongoing research efforts must analyze and monitor the ability of such initiatives to really make a difference in the global–local nexus of tourism development. In particular we must look carefully at the degree to which community power structures and inequalities may distort local commodification of tourist spaces, in effect making them no less able to meet the broader needs of local peoples than past, externally imposed constructions of tourist landscapes.

If the landscape of Canada's eastern Arctic is considered merely as a utilitarian object, a picture or a view, it will become captive to the frame through which it is viewed, both physically and culturally (Jacobs 1996). As Jacobs (1996: 72) notes:

> There is little hope and less purpose in freezing the north within a framework of traditions that no longer exist. But there is real meaning in illustrating these traditions as part of layered relationship of people and landscape.
>
> (Jacobs 1996: 72)

This chapter has clearly shown that the future of more appropriate tourism development in Canada's eastern Arctic lies, in large part, with the ability of its local people to construct "tourist places," and influence tourist perceptions and thoughts, in such a way that the industry works for them and meets local needs. We would argue that this is perhaps one of the central challenges facing the people of Nunavut as they prepare to enter the next millennium.

Notes

1   This research was funded by the Social Sciences and Humanities Research Council of Canada (grant #218–97 and 239–47), a Quebec FCAR Team Grant (#290–37) and several DIAND Northern Scientific Training Grants. The authors wish to acknowledge the critical inputs made by George Wenzel, and the research assistance provided by Rebecca Tarbotton.

References

Acres International Ltd (1988) *Baffin Visitors Survey*, Yellowknife NWT: GNWT.

Anderson, M. (1991) "Problems with tourism development in Canada's eastern Arctic," *Tourism Management*, 12 (Sept), 209–20.

Bell, C. (1996) *Inventing New Zealand: Everyday Myths of Pakeha Identity*, Auckland: Penguin.

Britton, R. A. (1979) "The image of the third world in tourism marketing," *Annals of Tourism Research*, 6 (3) 318–29.

Britton, S. G. (1982) "The political economy of tourism in the Third World," *Annals of Tourism Research*, 9 (3) 331–58.

Britton, S. G. (1991) "Tourism, capital and place: towards a critical geography of tourism," *Environment and planning D: Society and Space*, 9, 451–78.

Brody, H. (1991) *The People's Land: Inuit, Whites and the Eastern Arctic*, Vancouver BC: Douglas and McIntyre.

Campbell, C. (1995) "The sociology of consumption," in D. Miller (ed.) *Acknowledging Consumption*, London: Routledge, 96–126.

Canadian Broadcasting Commission (1992) *The Idea of North in Glenn Gould's Solitude Trilogy: Three Sound Documentaries*, CBC Records.

Cohen, E. (1993) "The study of tourist images of native people: mitigating the stereotype of a stereotype," in D. Pearce and R. Butler (eds) *Tourism Research: Critiques and Challenges*, London: Routledge.

Dann, G. (1996) "Images of destination people in travelogues," in R. Butler and T. Hinch, *Tourism and Indigenous Peoples*, London, International Thomson Business Press, 349–75.

Dilley, R. (1986) "Tourist brochures and tourist images," *Canadian Geographer*, 30 1, 59–65.

Dorfmann, J. (dir.) (1993) *Shadow of the Wolf*.

Flaherty, R. (dir.) (1922) *Nanook of the North*.

Friedman, M. (1996) "Northern exposure," *Netguide*, 3, 6, 61–4.

Government of the Northwest Territories (1983) "Community Based Tourism: A Strategy for the Northwest Territories," *Tourism Industry*, Yellowknife NWT: GNWT.

——(1991) *Tourism Marketing and Visitor Information Strategy, Baffin Island*, Iqaluit NWT: GNWT.

——(1992) *Baffin Region Tourism Industry Overview (1992)*, Iqaluit NWT: GNWT.

——(1993) *Marketing Strategy: Baffin region, 1993–1995*, Iqaluit NWT: Dept. of Economic Development and Tourism, GNWT.

Greenwald, B. (dir.) (1990) *Between Two Worlds*.

Greenwood, D. (1989) "Culture by the pound: an anthropological perspective on tourism as cultural commoditization," in V. Smith (ed.) *Hosts and Guests: The Anthropology of Tourism*, 2nd edn, Philadelphia PA: University of Pennsylvania Press, 171–85.

Grekin, J. (1994) "Towards an understanding of the community level impacts of ecotourism," unpublished M.A. thesis, Dept. of Geography, McGill University, Montreal.

Grekin, J. and Milne, S. (1994) *Tour Operators in the Northwest Territories*, McGill Tourism Research Group Industry Report no. 6, McGill University, Montreal.

——(1996) "Toward sustainable tourism development: the case of Pond Inlet, NWT," in R. W. Butler and T. Hinch (eds) *Tourism and Native Peoples*, London: International Thomson Business Press, 76–106.

Grise Fiord (1994) *Canada's Most Northerly Community*, community tourism brochure.

Hall, C. M. (1994) *Tourism and Politics: Policy, Power and Place*, Chichester: Wiley.

Hamburg, R. and Monteith, D. (1988) *A Strategy for Tourism Development in the Baffin Region*, Iqaluit NWT: Dept. of Economic Development and Tourism, GNWT.

Hamilton, R. W. (1993) *The Baffin Handbook: Traveling in Canada's Eastern Arctic*, Iqaluit NWT: Nortext.

Hamley, W. (1991) "Tourism in the Northwest Territories," *Geographical Review* 81 (4) 389–99.

Hinch, T. D. and Swinnerton, G. S. (1993) "Tourism and Canada's Northwest Territories: issues and prospects," *Tourism Recreation Research*, 18 (2) 23–31.

Hollinshead, K. (1992) " 'White' gaze, 'red' people—shadow visions: the disidentification of Indians in cultural tourism," *Leisure Studies*, 11, 43–64.

——(1996) "Marketing and metaphysical realism: the disidentification of aboriginal life and traditions through tourism," in R. W. Butler and T. Hinch (eds) *Tourism and Native Peoples*, London: Routledge, 308–48.

Hull, J. (1996) "Using the Internet to promote a sustainable tourism industry: a case study of the lower North Shore of Quebec," paper presented at the International Geographical Union, The Hague, 4–10 August.

Hummon, D. M. (1988) "Tourist worlds: tourist advertising, ritual and American culture," *The Sociological Quarterly*, 29 (2), 179–202.

Industry, Science, technology Canada (1990) *Federal Tourism Strategy: NWT*, Ottawa, Supply and Services Canada.

Jackson, P. (1989) *Maps of Meaning: an Introduction to Cultural Geography*, London: Unwin Hyman.

Jacobs, P. (1996) "The true north strong and free," *Ecodecision*, spring, 70–2

Johnston, M. (1995) "Patterns and issues in Arctic Tourism," in C. M. Hall and M. Johnson (eds) *Polar Tourism*, London: Belhaven Press.

Kaltenborn, B. (1995) "The value of polar wilderness in a global perspective," in V. Martin, and N. Tyler (eds) *Arctic Wilderness: the Fifth World Wilderness Congress*, Golden, North American Press, 103–10.

Keller, C. P. (1982) "The development of peripheral tourist destinations: case study, Baffin Region," unpublished M.A. thesis, Department of Geography, University of Western Ontario, London, Ontario.

Kulchyski, P. (1989) "The postmodern and the paleolithic: notes on technology and native community in the far north," *Canadian Journal of Political and Social Theory*, 13 (3), 49–62.

Lash, S. and Urry, J. (1994) *Economies of Signs and Space*, London: Sage.

Lopez, B. (1986) *Arctic Dreams: Imagination and Desire in a Northern Landscape*, New York: Scribner.

MacCannell, D. (1989) *The Tourist: A New Theory of the Leisure Class*, 2nd edn, New York: Schocken.

——(1992) *Empty Meeting Grounds: The Tourist Papers*, London: Routledge.

Milne, S. S. (1996) "Travel distribution technologies and the marketing of Pacific microstates," C. M. Hall and S. Page (eds) *Pacific Tourism*, London: Thomson International, 109–29.

Milne, S. S. (forthcoming) "Tourism and sustainable development: exploring the global-local nexus," in C. M. Hall and A. Lew (eds) *Tourism and Sustainable Development*, London: Routledge.

Milne, S. and Gill, K. (forthcoming) "Distribution technologies and destination development: myths and realities," in K. Debbage and D. Iaonnides (eds) *The Economic Geography of Tourism*, London: Routledge.

Milne, S. S., Ward, S. and Wenzel, G. (1995) "Linking Tourism and Art in Canada's Eastern Arctic: the case of Cape Dorset," *Polar Record*, 31 (176): 125–36.

Milne, S., Tarbotton, R., Woodley, S. and Wenzel, G. (1995) *Tourists to the Baffin Region: 1992 and 1993 Profiles*, McGill Tourism Research Group Industry Report no. 11, McGill University, Montreal.

Moss, J. (1994) *Enduring Dreams: An Exploration of Arctic Landscape*, Concord: Anansi House.

Mowat, F. (1976) *Canada's North Now: The Great Betrayal*, Toronto: McClelland and Stewart.

Nature Canada (1988a) letters to the editor, 17 (4), 5, 12.

Nickels, S., Milne, S. and Wenzel, G. (1991) "Resident perceptions of tourism development: the case of Clyde River, Baffin Island, NWT," *Inuit Studies*, 15 (1) 157–70.

"NWT Virtual Explorers' Guide" (http://www.edt.gov.nt.ca/guide/index.html).

Pelly, D. F. (1993) "Dawn of Nunavut: Inuit negotiate a home of their own in Canada's newest territory," *Canadian Geographic*, 113 (2) 20–9.

Preston, J. (1996) "Zapatista tour offers mud, sweat and radical chic," *New York Times*, 13 August, A4.

Purdy, A. (1967) *North of Summer: Poems from Baffin Island*, Montreal: McClelland and Stewart.

Reimer, G. (1989) "Resident Attitudes Towards Tourism Development in Pangnirtung, NWT," report prepared for the Northern Scientific training program, Hamilton, Ontario: McMaster University.

Reimer, G. and Dialla, A. (1992) "Community based tourism development in Pangnirtung, NWT: looking back and looking ahead," mimeo, Iqaluit NWT: GNWT.

Schama, S. (1996) *Landscape and Memory*, London: Fontana.

Silver, I. (1993) "Marketing authenticity in third world countries," *Annals of Tourism Research*, 20 2: 302–18.

Smith, S. L. J. (1994) "The tourism product," *Annals of Tourism Research*, 21, 582–95.

Squire, S. (1994a) "Accounting for cultural meanings: the interface between geography and tourism studies re-examined," *Progress in Human Geography*, 18 1–16.

——(1994b) "The cultural values of literary tourism," *Annals of Tourism Research*, 21, 103–20.

Urry, J. (1990) *The Tourist Gaze: Leisure and Travel in Contemporary Societies*, London: Sage.

——(1994) "Cultural change and contemporary tourism," *Leisure Studies*, 13, 233–8.

——(1995) *Consuming Places*, London: Routledge.

Wenzel, G. (1985) "Marooned in a blizzard of contradictions: Inuit and the anti-sealing movement," *Inuit Studies*, 9 (1) 77–91.

——(1991) *Animal Rights, Human Rights*, Toronto: University of Toronto Press.

Wilson, A. (1992) *The Culture of Nature: North American Landscape from Disney to Exxon Valdez*, Cambridge, Blackwell.

Woodley, S. (1996) "Ecotourism: opportunities and challenges for Nunavut," unpublished B.A. thesis, Department of Geography, Concordia University, Montreal.

7

TARTAN MYTHOLOGY

The traditional tourist image of Scotland

Richard W. Butler

Most successful tourism destinations of the modern era possess a selection of attributes which, although not identical from one location to another, tend to have certain characteristics in common. In general, the destinations are readily accessible to major markets, reasonable in cost, have good weather and possess a variety of activities and facilities of which the visitor can take advantage. Furthermore, they increasingly tend to possess heritage features, either cultural or natural or both, used as supplementary attractions. Indeed, it is relatively rare for locations to be successful in the long term as tourism destinations without such combinations of features. It is equally uncommon for tourism destinations to be able to maintain their attractivity over long periods of time unless they have unique and often spectacular features, such as Niagara Falls or the Pyramids, or are key locations such as the capital cities of London or Paris, or transportation hubs such as Singapore and Hong Kong.

It is perhaps puzzling, therefore, to imagine Scotland as a successful tourism destination over the long term. A small country on the periphery of Europe, long plagued by poor access and transportation, Scotland is costly compared to many other fringe areas, with a climate deemed unattractive to most visitors (and residents) and only limited facilities for tourists compared to other destinations. Yet the country has successfully attracted tourists for more than two hundred years as a destination, and to a considerable degree seems well positioned to continue attracting tourists over the long term.

To understand the paradox which Scotland represents as a tourist destination, it is necessary to interpret its origins and the image which it has presented and continues to present to potential visitors. Although this image is socially contrived, its origins are clear and well documented, and the attributes generally attached to Scotland over the past two centuries show little sign of declining in appeal. What has not been examined, however, is the relationship between past image creation, the evolution of tourists and the implications for Scotland's future as a tourist destination.

These features represent the focus of this chapter, which begins with a review of the image of Scotland as a destination, followed by an examination of the origins of that image. A discussion of the development over time and space of different attractions in Scotland, which reflect socio-economic changes in Britain, precedes an examination of how changes in tourism demand have changed the relative appeal of those attractions, and thus the overall attractivity of Scotland as a destination. The final section stresses the importance of the way in which Scotland is managed and marketed for tourism to different markets, and the implications for its future as a tourist destination within the vast and highly competitive array of global tourism destinations.

The image of Scotland

Compared to many countries in the world, Scotland has a very strong and distinctive image, even if this image is stereotypical and artificial to a great degree. In the course of teaching tourism courses over several years, this author has frequently surveyed the image held by students of selected countries. With reference to Scotland, the perceptual image has remained strongly consistent for two decades amongst this admittedly unscientific sample. It includes such primary features as mountains, tartan, bagpipes, castles and kilts, and secondary features such as highland dancing, haggis, heather, golf, Balmoral and lochs/lakes. Collectively, these descriptive features, as Gold and Gold (1995) note, provide the "shortbread" or "chocolate box" image so commonly associated with Scotland, a pastiche of scenes portrayed on containers of merchandise originating in or purported to be from or related to Scotland.

The importance of the overall image of Scotland as a country to the specific image of Scotland as a tourism destination is critical, yet often not addressed. While the desire of Scots, particularly those charged with marketing their country in the post-industrialized world, to portray the country as modern and ably equipped—particularly in its facilities, transportation and communications infrastructure—the appeal of Scotland to tourists appears to be very much tied to an image based on a limited and largely inaccurate historic picture very much at odds with the modern scene (Gold and Gold 1995). Therefore in portraying the modern Scotland promoters are, indirectly at least, threatening the very successful tourist image of Scotland, a point which is returned to later in this discussion.

The established image of Scotland which exists in many people's minds is a powerful one, and one which modern advertisers would have to work extremely hard to create. It has been reinforced for almost two centuries by a peculiar combination of geography, emigration, social stratification, romantic appeal and attachment in literature and art, and by the spatial segregation of different markets and development. To understand the origins of this image it is necessary to briefly explore the context in which it was first established.

Scotland as "terra horribilis"

In the eighteenth century, much of Scotland was viewed as a frightening wilderness by the rest of Britain and Europe. Three times in the first half of the century the mostly Catholic clans of the Highlands and Islands of northern and western Scotland had risen in rebellion against the Protestant Hanoverian royal family based in England, in efforts to restore the Stuart monarchy. The final rebellion of 1745–6 under Bonnie Prince Charlie was defeated at the battle of Culloden, and was followed by a period of military occupation and governance of much of Scotland (Prebble 1961). This occupation was undertaken in conjunction with a program of land confiscation and sale, language and dress restrictions, prohibition of weapons, and a range of other social and economic regulations aimed at permanently eliminating the social fabric which had supported and made possible the armed rebellions.

The feudal system of the Highlands was broken by force and by imposed social and economic changes, and was accompanied through the second half of the eighteenth century by extensive changes in the agricultural pattern of the region and subsequent forced emigration of much of the population. The "Highland Clearances" (Prebble 1963), which saw people removed and replaced by sheep, resulted in massive emigration to other parts of Britain, and particularly to Canada, Australia and New Zealand. The bulk of the population which remained was often relocated from the interior valleys (glens) to the coast, and traditional clan chiefs were replaced by absentee landlords, many of whom were from England or southern Scotland, and who subsequently allowed their estates to fall into disrepair.

These were often replaced in the nineteenth century by pseudo-castles and shooting lodges, many used for the summer periods only, frequently based in design and location on Rhineland castles and built in what is known as the Scottish baronial architectural style. Examples include Carbisdale Castle in Sutherland, the Trossachs Hotel in the Central Highlands, and perhaps above all, Balmoral Castle in Royal Deeside.

Not surprisingly, given its recent history of armed insurrection, Scotland was not seen as a holiday destination by anyone in the eighteenth century (Youngson 1974). Perhaps unexpectedly, however, the area did receive a large number of highly significant and noteworthy visitors who can be regarded quite justifiably as the precursors of later, more conventional tourists (Butler 1985). For the most part, these visitors fell into three distinct groups: those involved with the military occupation and administration, those engaged in scientific enquiry, and those involved with the world of letters, music and art.

The first group had relatively little direct importance on the development of tourism in the area or the establishment of the image of Scotland, but they did serve a distinguished role in furthering knowledge about Scotland, particularly the Highlands. People such as Burt (1754) wrote and published journals of their travels in the area, and while for the most part these publications stress the hardship of travel and life in the region, they did contain descriptions of the scenery and heritage that remained. They were also influential in persuading the occupying powers that improvements had to be made in transportation in the Highlands, if only to allow the occupying forces to move around more easily (Salmond 1934).

Scientific visitors were fewer in number initially, and most often were engineers and retired military or education officials intrigued with the geology and geomorphology of highland Scotland. They wrote widely and published in scientific journals as well as producing more literary works, and again served to publicize the distinctive and impressive physical features of Scotland, from Fingal's Cave (with basaltic columns similar to the Giant's Causeway of Ireland) to Ben Nevis (the highest mountain in Britain), and from Loch Ness to the Corrievreckan Whirlpool. They also frequently commented on and described much of the historical heritage of the area (Martin 1716; Penant 1772).

It is the third group of visitors, however, who had by far the greatest effect upon establishing the image of Scotland for tourism and in developing the fledgling tourism industry. This group consisted of authors especially, but also musicians, poets and artists. Indeed, a glance at the list of authors who visited Scotland between 1770 and 1870 reads like a *Who's Who* of English literature, including Defoe, Johnson and Boswell, Burns, Dickens, Coleridge, Wordsworth, Southey, Tennyson and Scott. What motivated them in the later years was essentially the romantic appeal of the area, but this appeal was in itself a partial creation of these visitors.

Defoe, one of the first of this literary group to visit this area, toured Scotland in the late eighteenth century as part of his tour of Great Britain, and described a visit to the Highlands as akin to a military expedition (1974), an appropriate term given the military presence there at that time. His writings were some of the first to document the attractiveness of the area, however, and set the tone for other visitors. Johnson and Boswell soon followed and their comments, some delightfully pithy, such as Johnson's on discovering that he had lost his walking stick on the Isle of Skye, when he remarked that this had contributed significantly to the amount of wood on the island, gave widespread publicity to the Highlands (Johnson 1775; Boswell 1952). Despite the tribulations of their travels, the reported exploits of these individuals did much to instill curiosity among prospective visitors to the area.

But it was the travel and writings of James MacPherson which truly placed highland Scotland and the Gael on the artistic and literary map of Europe in the eighteenth century. MacPherson produced and published what he claimed to be original poems in the epic style, by a Gaelic poet named Ossian (MacPherson 1765). These poems proved tremendously popular throughout Europe, and the image and exploits of the Gaelic hero Fingal and his cultural and spatial context attracted great curiosity. Ossian was reportedly the favorite poet of many European celebrities from Beethoven to Napoleon: "The combination of Celtic heroes, highland scenery, chivalry and fine emotions and sensibility . . . was an irresistible combination. It also led to new ways of seeing and representing Scottish landscape" (Gold and Gold 1995: 54).

This interest lasted for several decades and resulted in many visitors from Europe coming to the Highlands, including celebrities such as Mendelssohn, who composed the Hebridean Overture after visiting the Western Isles in 1820. Fingal's Cave, on the island of Staffa, is one of the few remaining pieces of evidence of this phenomenon, which disappeared quickly after MacPherson eventually admitted writing the poems himself, destroying the legend he had so effectively created (Gold and Gold 1995). The image of the Gaelic hero remained, however, and was to subsequently be re-established on an even larger scale by Sir Walter Scott a half-century later.

While Robert Burns, Scotland's national poet, also visited the Highlands, his travels were confined to a few areas and two visits only. He wrote several poems based on these two visits, but shared the lowland Scots' traditional apathy toward the Highlands and its inhabitants in general, a perspective perhaps based on an unfamiliarity with the landscape and customs, and memories of the "highland host" descending on the Lowlands to replenish their flocks and purses through the power of the sword. In reality, Burns and his visits did relatively little to promote the image of the Highlands for visitors; they had certainly nothing like the effect which his work did to enhance the appeal of his home region of southwest Scotland and the image of the rural lowland Scots who inhabited it.

One other group of visitors existed, and they were becoming more established. These were the absentee landowners and their friends, who had begun to visit their often vast estates in highland Scotland during the summer, and to partake in the traditional highland gentry's activities of hunting and fishing. The hunting of red deer in particular, and the fishing of salmon became prized attributes of these estates, many of which consisted of large expanses of moorland and mountain with very little agricultural value, even for those who introduced sheep, and whose forestry potential was not yet appreciated. However, the publication of several private journals revealed the wealth of such resources to a broader audience and clearly whetted the appetites of sportsmen in the south (Scrope 1847; Brander 1973).

Demand began to develop for access to such resources, and the principal of renting and leasing the shooting and fishing rights of some of the estates became established. The first commercial "let" of shooting property occurred in 1800 (O'Dell and Walton 1962: 332). As Orr (1982) pointed out, as this practice expanded, clearances of the population again took place, this time for deer and grouse, although on a smaller and more localized scale than those related to sheep ranching.

As a result of these actions, by the end of the eighteenth century Scotland had become a destination for a small and distinctive tourist market, but the country was still singularly ill-equipped to receive significant numbers of visitors. Access both to and within Scotland was poor, especially outside the central lowland belt, where it was notably difficult: "In the islands there were no roads, nor any marks by which a stranger may find his way" (Johnson 1775: 48). Consequently, land transport was poor to non-existent, except for a few military roads, and bridges rare, although considerably improved over earlier models, due to the efforts of Thomas Telford (Haldane 1962). While the Caledonian Canal bisected Scotland from southwest to northeast, its use was light and much of the Highlands remained relatively inaccessible. Visitor accommodations were similarly poor or unavailable, unless one was fortunate enough to have introductions to the landed gentry. Tourist travel thus took on the old meaning of travel, "travail," and those who made the effort normally had specific attractions which they wished to see and more than a passing interest in the region.

As the area gradually lost its threatening reputation as the home of a hostile population, as transportation slowly improved and inns began to be established, and as the amenity resources of the area became appreciated (Gilpin 1789), the scene was set for the total transformation of the region's image. The individual most responsible for this transformation was the novelist and poet Sir Walter Scott, appropriately known as the "Wizard of the North" for his ability to create imagery (Lockhart 1906). Scott virtually single-handedly changed the image of the Gael and his homeland from one of despair, unattractiveness, savagery and violence, to one of triumph, beauty, nobleness, and above all, romance. Through his poems and novels, Scott created and developed a mythology akin to that of King Arthur and the Round Table, one based on a mixture of reality and artistic licence, and presented in a style which perfectly caught the imagination of Victorian society.

Scott's books were bestsellers on a scale never witnessed before, and there is extensive evidence that his work had a phenomenal effect in changing the image of Scotland (Butler 1985; Durie 1996; Gold and Gold 1995). Perhaps his greatest single feat was to stage-manage the visit of George IV to Edinburgh in 1822, complete with clothing the king and his court in newly created tartan and pseudo-highland dress (Finlay 1981; Prebble 1989). This set the stamp of ultimate approval upon the idea of visiting Scotland and of wearing "highland dress" on ceremonial occasions. Scott's success has been discussed at length elsewhere (Butler 1973; Durie 1996; Gold and Gold 1995) and remains unparalleled in both the extent of transformation and its lasting effect: "He effectively wrote the script for the promotion of Scottish tourism through the nineteenth and twentieth centuries" (Gold and Gold 1995: 195). As a consequence, the image of the Scot became, and remains, the image of the tartaned highlander—even though highlanders have always represented a distinct minority in the history of Scotland (Trevor-Roper 1983).

Scott's efforts were reinforced both deliberately and coincidentally by his peers. Many authors and poets were attracted to Scotland by Scott's works, in particular Wordsworth, Coleridge, Southey and Dickens. Their own literary efforts also served to stimulate interest in Scotland and the Highlands in particular (Scott 1994). The great contemporary landscape artist Turner was engaged to illustrate editions of Scott's works, and these and other paintings he produced independently on visits to Scotland similarly evoked interest in the area. His artistic support of Scotland's image was reinforced by Landseer, Queen Victoria's favorite artist, who painted many Scottish scenes and backdrops for his work and others (Butler 1985).

Two other noteworthy features shaped the development of tourism in Scotland and its image in the early-to-middle nineteenth century, one relating to transportation, the other to royalty. The development of the railway system in Great Britain allowed access to many areas previously not easily visited. While the railway came late to the north and west of Scotland, steamship services were established early to the west coast and the islands, and served as the basis for much of the tourist travel (Butler 1973). Real growth in tourism came with the efforts of Thomas Cook, who began organized tours to Scotland in 1846, using rail as far north as possible and then transferring guests to steamers and stagecoaches for the remainder of the journey (Cook 1861). Participants in his tours visited many of the sites described in Scott's novels and poems, particularly the Trossachs, the central Highlands and the Isle of Skye, and Cook popularized the concept of the "Tour to the Highlands and Islands of Scotland," in many cases following closely the routes of Defoe, Johnson, Boswell and the early pioneers of "leisure" travel in the area.

The second noteworthy event was the visit by Queen Victoria to the Highlands in 1842, an act repeated in subsequent years and culminating in the purchase of the Balmoral property on Deeside and the subsequent construction of the present Balmoral Castle in 1855 (Brown 1955; Dudd 1983). The establishment of a royal summer holiday in the Scottish Highlands represented an endorsement of the concept of Scotland as a holiday destination of the highest social order. To the aristocracy, a Scottish estate became *de rigeur*, and tartans and tweeds, stalking rifles and salmon rods, and a Scottish "season" became an established part of the social order in Victorian society. For those unable to aspire to the requisite level of investment of money and time, Cook's tours and individual holidays presented an acceptable alternative. Regardless of the route it took, so influential was this perspective by the 1850s that Scotland was firmly established by the leading social figures as a respectable holiday destination.

Spatial and temporal patterns of destination development

A casual interpretation of the pattern of development of tourism destinations in Scotland might imply that this development had been deliberate and

127

planned. In fact most of the development followed, rather than preceded, the introduction of tourism to the areas concerned. It is equally important to note the distinction between travel and tourism, particularly in the early part of the period under discussion. There has always been travel to and within Scotland for a variety of purposes, but as stated above, very little of it was motivated by pleasure before the nineteenth century. Instead, the majority of travel was related to military matters, civil administration (particularly of justice) and agricultural transactions such as cattle droving (Haldane 1968). All of these elements, however, did play a role in stimulating the development of elements of the infrastructure necessary for pleasure travel or tourism, such as roads, bridges, ferry services and accommodation.

When the demands of tourism began to be experienced in the country, the initial response was one of improvement and adaptation of the existing infrastructure. The major development of facilities specifically designed for tourism followed a considerable time later. For the purposes of discussion, this section is divided into three very broad periods, and the spatial patterns of tourism discussed in the temporal context.

The pre-steam era (1750–1850)

Many of the key features of this period have been discussed above. Non-tourist travel during this era was focused on the principal towns (Edinburgh, Glasgow and Stirling in particular) for the purposes of administration and economic activity, and on military centers such as Fort William and Inverness during the period of military occupation. Ports, such as Leith and Aberdeen on the east coast and those on the Clyde in the west, saw considerable trade in goods, and were used for access to Scotland from other countries, as sea travel was both quicker and often more comfortable—even if a little more hazardous and unpredictable—than travel by very poor roads (Haldane 1962; Butler 1973).

Travel for leisure-related purposes, including social travel, and for scientific and artistic purposes as discussed above, was primarily directed into rural and peripheral areas. Unlike travel on the Grand Tour of Europe (Towner 1985), where the focus was on urban centers, capitals and other cultural attractions, eighteenth-century travel in Scotland was aimed at natural sites of interest and more isolated cultural features, many in the Highlands (Gilpin 1789). These included Fingal's Cave on Staffa and the Abbey on Iona (the site of the introduction of Christianity to Scotland), both off the west coast of Mull, and Dunvegan Castle on Skye, seat of the Clan Macleod and reputed to be the oldest occupied building in Scotland. Thus what might be described as tourist travel was primarily into areas little served by good roads or other services, and generally lacking accommodation and other facilities normally required by tourists. By necessity, such travel took considerable time and required thorough organizational skills and contacts. It is not surprising, therefore, that most of it was undertaken by those who possessed time, money and social links to the landed gentry (Hart-Davis 1974).

As travel increased following the enforced pacification of the Highlands, so did complaints about the lack of infrastructure and the difficulties of travel, complaints which were echoed for the next century or more by visitors (Burt 1754; Haldane 1962). The bridge construction and road improvement programs, undertaken mostly by the Army's Royal Engineers, allowed for easier road access, which along with military travel and the spread of the postal system throughout the area, encouraged the establishment of an increasing number of inns (Haldane 1962). As stagecoach services increased, so did the roles of these establishments, which evolved from simply providing basic shelter to the provision of food, accommodation and fresh horses, and which increasingly became social and economic foci for their communities (Butler 1973). During this period, however, hotels were almost entirely limited to the major towns, and even these were few in number and relatively primitive.

The major change affecting the development of tourism was related to the drastic changes in land ownership and the clearances discussed above, and in particular to the realization by absentee landlords that their often vast properties had great potential for sport. The early part of the nineteenth century saw the development of shooting lodges in many parts of the Highlands, and to a far lesser extent on the islands, and the rebuilding or renovation of the traditional homes of the gentry (O'Dell and Walton 1962; Orr 1982).

These gradually became used for hunting and fishing in the summer and autumn months, either by their owners and friends, or by others to whom they were rented or leased. The difficulty and expense in time and money of reaching these properties meant that they were inhabited fairly continuously by all or some of the family for the summer only, and closed by the mid-autumn. Figures do not exist on the total number of such properties extant in the early part of the nineteenth century, but there is no doubt that they increased rapidly following the state visit of King George to Edinburgh in 1822.

The age of steam (1850–1930)

There are two dimensions to this period, one relating to the coasts, the other to the interior of Scotland. The appearance of the steamship allowed much more reliable, safer and to some extent faster transportation to and around the coasts of Scotland. Tourists, as well as other travelers, were quick to take advantage of this innovation (Anonymous 1836). In the initial period of its introduction, the steamship allowed the expansion of tourism to already popular areas—the Western Isles and the east coast, for example Fife (Anonymous 1830)—and to reach, in small numbers, the Northern Isles of Orkney and Shetland (Butler 1996). Steamships also allowed Thomas Cook to introduce mass tourism to parts of Scotland in the years before rail reached the country, by transporting tourists by train to Liverpool and to other gateway ports, and then by ship to

Ardrossan and Glasgow. On arrival in Scotland, the tourists were moved by stagecoach and steamer to their final destinations (Cook 1861).

The introduction of the steamship also facilitated the development of the only real mass tourism development witnessed in Scotland for almost a century, namely the creation of a series of tourist resorts on the Firth of Clyde, some 80 to 150 kilometers from Glasgow and the Clyde Valley industrial towns. These centers—including Helensburgh, Gourock, Ardrossan, Ayr and Largs, all served later by rail, and Dunoon, Innellan, Rothesay and Brodick, served solely by steamers—grew rapidly through the nineteenth century. Together, they served the relatively captive market of the industrial belt of western Scotland, whose citizens would have faced considerable time and expense to reach other holiday centers (Pattison 1967). Indeed, so strong a monopoly did the communities hold on this visitor population that they continued to attract large numbers of tourists well into the second half of the twentieth century, long after their attraction had declined relative to resorts in other parts of Europe and the world. Due largely to their proximity and accessibility, the dated attractions remained viable as tourist destinations until the 1960s, and many still attract a fair proportion of the elderly population from their original travel market.

The railways both accentuated these developments and helped create new developments. On the east coast, similar but smaller developments occurred at Joppa and Musselburgh near Edinburgh. The railway companies joined forces with the steamship companies, in some cases merging, and offered cheap and convenient travel from home to holiday resort on the Clyde estuary. They also broadened the market catchment area to include urban centers in northern and central England, especially those centers whose industrial holidays were scheduled at different times to those of the Clydeside urban centers. They further invested in hotels, piers, amusements and other facilities, and competition between companies and individual boats was fierce (Paterson 1969; McCrorie 1986). While these destinations have since switched in market focus, many remain primarily dependent on sea access, albeit in a different form.

The railways also provided access to selected areas of the Highlands, in particular those areas adjoining the rail routes. The nature of Scotland's topography meant that the rail routes were few and late in coming, in some cases not being completed until the twentieth century, over fifty years after the railway first reached Scotland. Towns such as Oban, "the headquarters of all who desire to visit the western highlands" (Groome 1894: vol. 2, 124) boomed with the introduction of the railway and became a railhead of great importance, connecting to a fleet of steamers for tourist and local traffic. Mallaig and Kyle of Lochalsh benefited similarly on the west coast, and towns such as Aberdeen, Inverness and Thurso also grew rapidly, although their growth was more directly stimulated by non-tourist development.

The railway did provide stimulus for tourism in some other specific localities. The Spey Valley was one such area, and the small towns of Newtonmore, Kingussie, Aviemore, Boat of Garten and Grantown-on-Spey all witnessed local booms in hotel and guest house construction as the railway increased their accessibility. The salmon in the Spey River, walks in the Cairngorms and Monadhliath Mountains, and the shooting of grouse and red deer made Speyside a close rival to Deeside in its attractiveness to wealthy tourists. A smaller group of resorts developed north of Inverness, and a spa was created at Strathpeffer, although the local landowner prevented the railway from actually going through the village (Butler 1973).

Royal Deeside also benefited greatly from the introduction of steam travel (Farr 1968). The royal estate at Balmoral could now be reached almost entirely by rail from London, via Aberdeen and the new Forth and Tay bridges, as the Deeside line reached as far as Ballater, only seventeen miles from Balmoral. A few isolated shooting lodges were also made more accessible by the railway, and some peculiar and distinctive arrangements over routing and service were extorted by the estate owners to either ensure service or prevent disturbance by railways on their properties (Vallance 1972).

Consequently, the advent of steam-powered travel permanently altered the face of Scotland as a tourist destination. In the first place, it re-emphasized the appeal of the Highlands and Islands by drastically improving the speed, comfort and safety of access to these areas, or at least parts of them (Thomas 1965). As such, it allowed specific areas to grow at accelerated rates compared to areas without such access. Second, it opened up some new areas of the Highlands, such as the Spey Valley, the Northern Isles, the Moray Firth Coast east of Inverness, and Deeside. In a totally different vein, it allowed and even actively encouraged the development of true mass tourism in the Firth of Clyde and a few other areas east of Edinburgh, and in the Dornoch Firth.

Furthermore, it made possible the introduction of a second form of mass tourism in terms of tours of the type organized by Thomas Cook, who was particularly adept at coordinating various forms of travel-related services to proffer opportunities unknown before his innovations (Swinglehurst 1974). Cook certainly continued to market Scotland on the basis of the appeal created by Sir Walter Scott, but soon broadened the geographical base of his tours to other areas as he saw their potential for attracting visitors (Cook 1861). Scotland thus became not one generalized destination based on a vague Celtic mythical image, but a series of destinations, often clearly defined and aggressively marketed to specific markets. In addition, the railway provided the only means of access for a significant element of the population who wished to walk and climb on the Scottish mountains.

Laws of access in Scotland allow freedom from trespass charges except in cases of damage to property, and in the inter-war years extensive use was made of the hills within a day's journey of the major urban centers. Areas such as Loch Lomondside, the Trossachs and the Angus Glens were easily reached by special "walkers' trains" which continued until 1957 (Thomas 1965), while the establishment of the Scottish Youth Hostels Association, and the development of their chain of hostels offering budget accommodation in many of the most popular areas, enabled large numbers of less affluent visitors to access such areas for longer periods for recreation and tourism. "Mass" tourism in the rural areas of Scotland, therefore, had its beginnings in this period when it was first made possible by rail, but it was soon dominated by road transport, particularly the automobile.

The age of the automobile (1930–present)

In many respects much of the history of tourism in this period is unknown. The development of the steamships and railway era is well documented, and tourism in the decades from the 1960s onwards has been the subject of modern academic interest (see Seaton and Bennett 1996), but the period from the end of the nineteenth century to the Second World War is relatively unstudied. As the automobile became popular in Britain (noticeably later than in North America) so roads began to be improved and tourists shifted to this mode of transport in lieu of rail. Yet Scotland remained disadvantaged as a tourist destination in the automobile age.

It lay relatively far from the major markets of Europe and even England, and motorway construction in Britain did not begin until the 1960s. Even in 1997 there is still no continuous motorway link between Scotland and England, a continuing deterrent to motorists who desire expedient travel to their destinations.

For much of the prewar period, the tourist areas of Scotland were still served primarily by rail and/or steamship (McCrorie 1986) and the Clyde Coast resorts continued to host a considerable fleet of steamers until the late 1960s, when they were replaced by less attractive but more functional roll-on-roll-off car ferries—a reflection of the changed behavior and nature of many tourists. The massive reduction in rail service in Britain in the 1960s eliminated access to many rural areas, although in relative terms Scotland was not as hard-hit as other parts of Britain. Instead, the continuation of many rural services allowed tourism to continue by rail much later than in other areas of Britain, and in some ways delayed changes in the status quo of Scotland's tourism markets and activities, as rail access remained available for perhaps two decades longer than elsewhere in Britain.

Travel by private and commercial vehicle to Scotland really began in the 1920s, resulting in the emergence of a new form of literature geared to the touring motorist, as illustrated by the works of Morton (1929, 1933), and a wider market (Lambert 1997: personal communication). Already, automobiles and motorcoaches transport 65 percent of all tourist arrivals (Scottish Tourist Board 1995), a position that has remained unchanged for the last quarter-century (Butler 1973). Meanwhile, ongoing improvements to roads and bridges in Scotland since the 1960s assure that travel by private and commercial vehicle will continue to be the predominant form of tourism in Scotland. Indeed, many of the large hotels built to serve the railway trade would have faced a very unprofitable existence, and even bankruptcy, had they not been filled by coachloads of middle-class tourists on "scenic Highland tours" in the immediate postwar years (Butler 1973). British coaches are now matched with ever-increasing numbers of continental coaches making similar tours.

However, the routes remain little varied, since construction of new roads (as compared to road improvements) has been extremely limited in Scotland. In the Highlands, probably fewer than fifty miles of new road, i.e. roads serving new routes, have been developed in the last century. Thus the sites seen and the places visited are still those explored by earlier tourists nearly two hundred years previously. As will be discussed below, then, it is only in the context of newly evolving forms of tourism that fresh spatial patterns have emerged and new destinations within Scotland developed.

The most common pattern of travel is for visits to Edinburgh, and increasingly to Glasgow, with a circuit of the Highlands and possibly some of the islands, lasting between one and two weeks, and most stays limited to overnight stopovers. As a result, the old resorts on the Firth of Clyde and elsewhere have lost most of their appeal for long-stay family holidays and have now adopted, in response, a variety of ancillary roles, including service as commuting centers (particularly those on the mainland with good rail access to Clydeside), retirement havens, seasonal home communities, and centers for special-interest tourism activities such as yachting, golf and adventure tourism.

New bridges constructed across the Forth and Clyde rivers in the 1960s also offered easier access north to many tourists. The replacement of ferries with bridges, especially at Ballachulish and Kyle of Lochalsh, simultaneously reduced or removed major traffic impediments, although the Skye bridge was not welcomed by locals who mourned the loss of their "islandness" and the projected increase in traffic and other social disturbances on Sundays.

There have also been significant improvements in air transportation, with increased and improved service particularly to the Western and Northern Isles. While the improvement of these services is stimulated primarily by oil exploration and other industrial development rather than tourism demand, particularly in the case of Orkney and Shetland, the effect on tourism has been positive (Butler and Fennell 1994). While a dramatic increase in visitor numbers has not yet materialized, the cumulative effect of the transportation improvements has been the extension of the outer limits of tourism further north and west in Scotland.

It is now possible to reach the outer islands or the northwest of the Scottish mainland in a few hours by car from major urban centers, thereby making all of Scotland perceptually acceptable for weekend trips by residents of the Central Lowlands, where the bulk of the Scottish population resides. With high-speed (by British standards) train access from Europe to Glasgow and Edinburgh, vastly improved air service from English and continental gateway airports, and the new breed of roll-on-roll-off car ferries, all of Scotland is now perceived as accessible and equally suited for either extended holidays or seasonal residency.

The overall attractivity of Scotland as a destination

In the last twenty-five years the overall attraction of Scotland as a destination has increased significantly, despite significant competition from similarly situated destinations. Given the depiction of Scotland described earlier in this chapter, such a rate of growth may appear paradoxical, but the reason has its origins in two main points. First, the cultural and natural heritage of Scotland is still perceived by tourists as genuine, and therefore authentic and attractive at a time when much of the tourist world's heritage is blatantly artificial and staged (a point elaborated on in the conclusion to this chapter). The second reason is that the elements associated with Scotland's attractiveness as a destination have increased in quality and variety.

Some of the reasons behind this improvement in quality are related to external forces, including the provision of financing from the European Union and the requirements of the oil industry (Butler and Fennell 1994). In other cases, however, it reflects increased optimism among companies involved in tourism regarding its future in Scotland. For most, these investments are long-overdue and are simply bringing Scottish standards of accommodation and infrastructure up to industry standards, but they are improvements nevertheless.

It is the increase in the diversity and variety of attractions which has made the most difference in overall attraction and appeal to tourists. Nearly all of the key features relate to either the cultural and/or natural heritage of Scotland, and each reflects and capitalizes on the evolving changes in global tastes and preferences of tourists. Within the context of Scotland's natural environment, the development of skiing in Scotland, which began on an economic basis in the Spey Valley and Glencoe in the 1960s, has engendered significant impacts, not only in terms of developing a second season in these areas and also Deeside and Fort William in later years, but just as important, in stimulating improvements to accommodations and the construction of additional infrastructure. Likewise, the successful development of skiing provided added publicity for the regions and served to encourage private entrepreneurs to develop related facilities and opportunities for summer visitors also.

There has been a significant increase in participation in other outdoor activities beside skiing, including water sports, natural history, walking and adventure tourism, and entrepreneurs are now actively promoting such opportunities, including instruction and training in such activities. One of the earliest examples is pony-trekking, which originated in Scotland in Newtonmore in the Spey Valley in the early 1960s, and has now spread to many other areas with considerable success. As a result of these and similar efforts, Scotland—especially the rural mountainous areas—is increasingly and successfully marketed for a variety of activity holidays, tapping a new and different market to the traditional car-borne, middle-aged tourists of the first half of the twentieth century or the families from the Central Lowlands journeying to the Clyde Coast resorts.

In the context of cultural heritage there has been less true innovation. While culture has always been one of the primary attractions of Scotland, as noted above, only in the last quarter-century has the cultural heritage been more skillfully and aggressively marketed (Gold and Gold 1995; McNiven 1994). Heritage properties have been opened in greater numbers and interpretative efforts vastly improved, along with many more opportunities for extracting money from tourists. Commercial pony-trekking on the royal estate at Balmoral is now possible, and the royal tea rooms and gift shop are mirrored in most stately homes in Scotland.

Single-purpose interpretative centers have met with equal success as tourist attractions, ranging from the excellent products of the National Trust for Scotland to the Loch Ness Monster Centre and several whisky-tasting centers. The incorporation of scotch into the tourist appeal of Scotland is an excellent example of the use of heritage, and has been well described by McBoyle (1994). Employment as guides at distilleries probably now exceeds employment related to the production of whisky at these distilleries, while "whisky trails" now successfully compete with "fishing trails" and other similar marketing devices intended to persuade the tourist to visit specific locations.

Marketing other examples of Scotland's cultural heritage is also increasingly lucrative. Edinburgh has long been internationally famous for its festival, held at the end of the summer, and it now appears to be attracting similar numbers over the last two years for its new-year (hogmanay) celebrations. Glasgow, long eclipsed by the cultural attractions of its urban rival, scored a remarkable coup in the 1980s by being named Europe's "City of Culture." The effect on the city has been significant, resulting in the cleaning and display of the city's magnificent collection of eighteenth- and nineteenth-century architecture, and a marked increase in visitation as the entire image of the city has risen spectacularly over the last decade. Thus not only has the quality of Scotland as a destination increased but so too has its variety. While the "shortbread tin" image remains fundamental, there is now a much broader base to its appeal.

Conclusions

The preceding discussion has emphasized Scotland's cultural and natural heritage as the base of its appeal as a tourist destination. Scotland must exist as a specific destination on the tourist stage, for it is too peripheral and distant from major transit routes to draw tourists en route to other destinations. Indeed, that it has managed to remain attractive to tourists for the last two hundred years is rather remarkable, given these formidable impediments and the relative paucity of attractions compared to many other destinations. Yet over the past decade the number of visitors has remained relatively consistent (Scottish Tourist Board 1995). Thus the key question now is whether Scotland will be able to maintain this attractivity in the future. The answer depends on how it is marketed and managed, and how the attractions and heritage are maintained so as to retain their integrity and appeal.

Undoubtedly, the Scots themselves have a great deal to do with the strength of Scotland's tourist image. The widely flung Scottish diaspora represents a recognizable group that often preserves its Scottishness at a level far higher than is maintained in Scotland. There are more cohesive and active Burns Clubs, St Andrews Clubs and Highland Games in Canada, the USA, Australia and New Zealand, than are found in Scotland. Added to these associations are the Scottish sports teams whose performance, though not always successful, serves to remind the world of the Scottish presence in soccer and rugby in particular. Scotland as a national entity, therefore, gains recognition and image endorsement globally, even if that image is often the tartaned highlander-with-bagpipes of Sir Walter Scott.

While this author does not disagree with the conclusions of Gold and Gold (1995: 202) who state that "conventional promotions and policy propagate a conservative and incomplete picture of Scotland . . . [and] may now limit the potential of tourism rather than expand it," there is a real danger that abandonment of the stereotyped image of Scotland will also limit tourism. It is correct to argue that, if all that is marketed is "Scott's Scotland," such a policy would fail to capitalize on all of Scotland's assets. Missed opportunities already abound—the lack of emphasis on and protection for industrial heritage resources provides one example—but the fact remains that the tartan image is a very strong and positive one which has been successful for two centuries.

To discredit this image or to significantly downplay it, because it may be politically or historically incorrect or incomplete as an image for the country as a whole, could be a mistake of Ossianic proportions itself. The tartan image is uniquely associated with Scotland and highly recognizable. Any alternative, while perhaps more contemporary and accurate, is less likely to be as popular and successful over such a long period. Thus, combining contemporary elements and attributes with the established image would appear to be the most realistic option in marketing Scotland as a destination in the future (the continuing revamping of one brand of Scotch's famous advertising slogan "Born 1820, still going strong" would seem an appropriate analogy here).

Other trends also suggest that the future for Scottish tourism may continue in a positive vein. The rise in participation in activity holidays and those that are nature-related has been noted above, and Scotland is well situated geographically in Europe to benefit from such trends, if for no other reason than the limited economic options for much of the landscape and the unlikely possibility of change in the future. Indeed, tourism and recreation may constitute the most significant threat to many of the most popular and most visited areas (Dickinson 1996). Other, more traditional Scottish activities are also markedly increasing in popularity, with golf perhaps the best example. The self-proclaimed "Home of Golf," and exceptionally well endowed with good facilities, Scotland stands well positioned to gain from its premier reputation and the tremendous growth in this sport alone.

In recent years Scotland has also benefited, as has Australia, from successful movies and television productions (Riley 1994). Just as Australia had *Crocodile Dundee*, *The Man from Snowy River*, and even *Neighbours*, so Scotland has *Braveheart*, *Rob Roy* and a number of television programs now syndicated to a broader English-language audience. In this manner the Scotland portrayed via celluloid and satellite may build on, and perhaps modify, the image of Scotland depicted on paper and in print for the last two centuries. But old images are hard to shake (Butler 1996), and just as salmon, venison, shortbread and whisky have been gastronomic favorites for many years, it is likely that tartan, bagpipes, castles and mountains will remain mental and aesthetic favorites for some time to come.

References

Anonymous (1830) *The Steam Boat Companion; or a Stranger's Guide to the Western Highlands and Islands of Scotland, including Staffa, Iona and Other Places Usually Visited by Travellers; with a Topographical Description of the River Clyde and the Adjoining Scenery of Loch Lomond*, Glasgow: James Lumsden and Son.

——(1836) *The Scottish Tourist*, Edinburgh: Stirling, Kenney.

Boswell, J. (1852) *Journal of a Tour to the Hebrides with Samual Johnson*, London: National Illustrated Library.

Brander, M. (1973) *A Hunt Around the Highlands*, London: The Standfast Press.

Brown, I. (1955) *Balmoral, The History of a Home*, Glasgow: Collins.

Burt, E. (1754) *Letters from a Gentleman in the North of Scotland*, Edinburgh: Donaldson.

Butler, R. W. (1973) "The tourist industry of the Highlands and Islands," unpublished Ph.D. thesis, University of Glasgow.

——(1985) "Evolution of tourism in the Scottish Highlands," *Annals of Tourism Research*, 12, 2: 371–91.

——(1996) "Tourism in the Northern Isles: Orkney and Shetland," in D. G. Lockhart and D. Drakakis-Smith (eds) *Island Tourism*, London: Pinter, 59–80.

Butler, R. W., and Fennell, D. A. (1994) "The effects of North Sea oil development on the development of tourism," *Tourism Management*, 15, 5: 347–57.

Cook, T. (1861) *Cook's Scottish Tourist Official Directory*, Leicester: T. Cook.

Defoe, D. (1974) *A Tour Through the Whole Island of Great Britain*, London: Dent.

Dickinson, G. (1996) "Environmental Degradation in the Countryside: Loch Lomond, Scotland," in Priestley, G., Edwards, J. A. and Coccossis, H. (eds) *Sustainable Tourism? European Experiences*, CAB International, Wallingford, 22–34.

Dudd, D. (ed.) (1983) *Queen Victoria's Highland Journals*, Exeter: Webb and Bower.

Durie, A. J. (1996) *The British Linen Company, 1745–1775*, Edinburgh: Phillans & Wilson.

Farr, A. D. (1968) *The Royal Deeside Line*, Newton Abbot: David and Charles.

Finlay, G. (1981) *Turner and George the Fourth in Edinburgh in 1822*, London: Tate Gallery and Edinburgh University Press.

Fontane, F. (1965) *Across the Tweed*, London: Phoenix House.

Gilpin, W. (1789) *Observations Relative Chiefly to Picturesque Beauty, Made in the Year 1776 on Several Parts of Great Britain; Particularly the High-Lands of Scotland*, vol. 1, London: n.p.

Gold, J. R. and Gold, M. M. (1995) *Imagining Scotland*, Aldershot: Gower Press.

Groome, F. H. (1894) *Ordinance Gazetteer of Scotland*, London: Mackenzie.

Haldane, A. R. B. (1962) *New Ways Through the Glens*, London: Thomas and Nelson.

——(1968) *The Drove Roads of Scotland*, London: Thomas and Nelson.

Hart-Davis, D. (1974) *Peter Fleming: A Biography*, London: Cape.

Hughes, G. (1992) "Tourism and the geographical imagination," *Leisure Studies*, 11, 1: 31–42.

Johnson, S. (1775) *A Journey to the Western Islands of Scotland in 1973*, London.

Lockhart, J. G. (1906) *The Life of Sir Walter Scott*, London: Dent.

McBoyle, G. (1994) "Industry's contribution to Scottish tourism: the example of malt whisky distilleries," in A. V. Seaton (ed.) *Tourism: State of the Art*, Chichester: John Wiley and Sons Ltd, 517–28.

McCrorie, I. (1986) *Clyde Pleasure Steamers: An Illustrated History*, Greenloch: Orr, Pollock.

McNiven, D. (1994) "Presenting historic Scotland," in J. M. Fladmark (ed.) *Cultural Tourism*, Wimbledon: Donhead, 225–36.

MacPherson, J. (1765) *The Works of Ossian, the Song of Fingal*, trans. James MacPherson, Edinburgh: n.p.

Martin, M. (1716) *A Description of the Western Islands of Scotland*, Edinburgh: Mercat Press.

Morton, H. V. (1929) *In Search of Scotland*, London: Methuen.

——(1933) *In Scotland Again*, London: Methuen.

O'Dell, A. C. and Walton, K. (1962) *The Highlands and Islands of Scotland*, London: T. Nelson and Sons.

Orr, W. (1982) *Deer Forests, Landlords and Crofters: the Western Highlands in Victorian and Edwardian Times*, Edinburgh: John Donald.

Paterson, A. J. S. (1969) *The Golden Years of the Clyde Steamers*, Newton Abott: David and Charles.

Pattison, D. A. (1967) "Tourism in the Firth of Clyde," unpublished Ph.D. thesis, University of Glasgow.

Penant, T. (1772) *A Tour in Scotland, and A Voyage to the Hebrides*, Chester: n.p.

Prebble, J. (1961) *Culloden*, London: Secker and Warburg.

——(1963) *The Highland Clearances*, London: Secker and Warburg.

——(1989) *The King's Jaunt: George IV in Scotland 1822, "One and Twenty Daft Days,"* London: Fontana.

Riley, R. W.(1994) "Movie-induced tourism," in A. V. Seaton (ed.) *Tourism: State of the Art*, Chichester: John Wiley and Sons Ltd, 453–8.

Salmond, J. E. (1934) *Wade in Scotland*, Edinburgh: Moray Press.

Scott, P. H. (1994) "The image of Scotland in literature," in J. M. Fladmark (ed.) *Cultural Tourism*, Wimbledon: Donhead, 362–73.

Scottish Tourist Board (1995) *Tourism in Scotland 1994*, Edinburgh: STB.

——(1996) *Scotland: Where to Go and What to See*, Edinburgh: STB.

Scrope, W. (1847) *Days of Deer Stalking in the Forest of Atholl*, London: John Murray.

Seaton, A. V. and Bennett, M. M. (1996) *The Marketing of Tourism Products: concepts, issues, and cases*, London: International Thomas Business Press.

Swinglehurst, E. (1974) *The Romantic Journey*, New York: Harper & Row.

Thomas, J. (1965) *The West Highland Railway*, Newton Abbott: David and Charles.

Thornton, T. (1896) *A Sporting Tour Through the Northern Parts of England, and a Great Part of the Highlands of Scotland, Including Remarks on English and Scottish Landscapes and General Observations on the State of Society and Manners*, London: Sportsman's Library.

Towner, J. (1985) "The grand tour: a key phase in the history of tourism," *Annals of Tourism Research*, 12, 2: 297–333.

Trevor-Roper, H. (1983) "The invention of tradition: the Highland tradition of Scotland," in E. Hobsbawm and T. Ranger (eds) *The Invention of Tradition*, Cambridge: Cambridge University Press.

Vallance, H. A. (1972) *The Highland Railway*, Newton Abbott: David and Charles.

Youngson, A. J. (1974) *Beyond the Highland Line*, London: Collins.

8

MAKING THE PACIFIC

Globalization, modernity and myth

C. Michael Hall

In the minds of many Western tourists the idea of the Pacific conjures up impressions of swaying tropical palm trees, white sand beaches, warm, crystal-clear waters and, possibly, dusky maidens in grass skirts or sarongs. This stereotypical and highly gendered image of "paradise" has been consistently portrayed over many years, not only in tourist advertising but also in many other forms of image making, such as film, newspapers and magazines, novels and even academic works. Such images are an inherent part of the tourism phenomenon which, perhaps more than any other business, is based on the production, reproduction and reinforcement of images.

Consequently, these conceptions of tourist places serve to project the "other" into the lives of consumers and, if successful, will assist in setting the socially constructed boundaries of a network of attractions referred to as "destination." Otherness is an essential component of tourism, for "[e]ncounters with the 'other' have always provided fuel for myths and mythical language. Contemporary tourism has developed its own promotional lexicon and repertoire of myths" (Selwyn 1993: 136). For the vast majority of people, otherness makes the destination attractive for consumption by establishing its distinctiveness.

Ironically, though, while "large numbers of tourists may be attracted to the region by its perceived 'differentness,' lured by the images of culture and land-scape which are vividly portrayed in the promotional literature, few are able or willing to tolerate a great deal of novelty" (Hitchcock *et al.* 1993: 3). At the same time, the process of "producing" cultural landscapes for tourist consumption makes one dependent on the other, for there can be no consumption without production: "It is apparent that they merge in many places and that each process certainly does have effects on the others even if they are causal or may never ever be explicable" (Laurier 1993: 272). Any meaningful understanding of the creation of the destination, therefore, involves situating the artfully constructed representation of that destination within the context of place consumption and production, and more particularly, the manner by which places are incorporated into the global capital system.

As a destination for the tourist, the Pacific is a creation of capitalism. While the Pacific Ocean delineates the physical boundaries of the Pacific region, it is the socially constructed Pacific with its attendant myths which dominates the tourist's mind and is commodified for the tourist's pleasure (and capitals, which is by far the most important). However, place-making is not just a tourism phenomenon, for the region's identity also reflects the means by which the Pacific was—and continues to be—incorporated into the global capitalist system.

The Pacific and global capitalism

Certainly, the Pacific region has always been a component of the global system (though the process of globalization itself has long been contested, as evidenced by disagreements between those who adopt a "Marxist" approach in their support of Wallerstein's World Systems Theory *v.* those who take the more conventional, "liberal" Robertson version of events). Migrations and trading relations have long linked different peoples throughout the region and the wider world. However, prior to the introduction of a capitalist economy, multiple centers of power and dominance existed. At the same time, the rate of growth and interaction in global trade and, arguably, cultural exchange, was considerably slower than that which obtains today. It is this particular set of relationships and links, and the corresponding increase in the rate of exchange of capital and culture, now termed globalization, which is of most interest (Bell 1979; Tomlinson 1991; Friedman 1994).

While Hoogvelt (1982) asserts that imperialism coincided with capitalism, other authors contend that the emergence of capitalism in Europe in the Middle Ages coincided with the imperialist yearnings of the European powers, improvements in transport technology and the development of a mercantile class in search of raw resources, produce and trade with the ever-widening known world (e.g. Tawney 1938; Agnew 1987; Knox and Agnew 1989). Exploration became a geographical activity driven by the urgencies of economic growth, and the "discovery" of the Pacific by Europeans became a crucial point in defining the image of the Pacific that prevails today.

Early trading with India and the Spice Islands of the Indonesian archipelago served to stimulate a sense of the "exotic," a feeling enhanced by the French and English voyages of the seventeenth and eighteenth centuries which confirmed the discovery of "paradise." Contributing to this picture were two factors strongly influencing the Western mind in this period: the writings of Jean-Jacques Rousseau (1978) and the reassessment of Classicism, which had been stimulated by the unearthing of Herculaneum and Pompeii (Honour 1981). It was in islands of the Pacific that Rousseau's romantic "noble savage," elements of which had already been identified in the peoples of

141

the Americas and Southeast Asia, was to be discovered. When the French explorer Louis Antoine de Bougainville landed in Tahiti in April 1768, he marveled how nature had endowed the island and allowed a people to live in happiness barely removed from the state of nature. Sentiment rather than reason vanquished unpleasant aspects of his stay, and in departing, he stated his desire to forever "extol the happy isle of Cythera: it is the true Utopia" (Brown 1988: 12).

Bougainville was not alone in the sentiments he expressed. In the English-speaking world, publications of various accounts from Cook's voyages served to further establish the Pacific's romantic image, as did Joseph Banks' use of Greek heroes when naming the inhabitants of Tahiti (Smith 1960). If Bougainville cherished visions of an "Island of Love," Tahiti became, for Banks, the "truest picture of Arcadia" (Brown 1988: 12).

However, the romantic picture of the Pacific created in Tahiti was not restricted to the island itself. Instead, Sydney Parkinson, a natural historian and artist in New Zealand, described Tolaga Bay in similarly glowing terms:

> The country about the bay is agreeable beyond description and, with proper cultivation, might be rendered a second Paradise. The hills are covered with beautiful flowering shrubs, intermingled with a great number of tall and stately palms, which fill the air with the most fragrant perfume.
>
> (Brown 1988: 13)

Meanwhile, similar perceptions were expressed by visitors to Australia. For example, early commentators on the area upon which Sydney is now situated noted (in Seddon 1976: 10):

> The general face of the country is certainly pleasing, being diversified with gentle ascents, and little winding valleys, covered for the most part with large spreading trees, which afford a succession of leaves in all seasons. In those places where trees are scarce, a variety of flow-ering shrubs abound, most of them entirely new to a European, and surpassing in beauty, fragrance and number, all I ever saw in an uncultivated state: among these, a tall shrub, bearing an elegant white flower, which smells like English May, is particularly delightful, and perfumes the air to a great distance.
>
> (Captain Watkin Tench)

> The greater part of the country is like an English park, and the trees give it the appearance of a wilderness or shrubbery, commonly attached to the habitations of people of fortune, filled with a variety of native plants, placed in a wild, irregular manner.
>
> (Mrs Elizabeth MacArthur)

To describe the beautiful and novel appearance of the different coves and islands as we sailed up is a task I shall not undertake, as I am conscious I cannot do justice to the subject. Suffice it to say that the finest terra's, lawns and grottos, with distinct plantations of the tallest and most stately trees I ever saw in any nobleman's grounds in England, cannot excel in beauty those which nature now presented to our view.

(Surgeon Bowes)

Admittedly, such perceptions were not held by all. The new inhabitants of Australia were faced with a bizarre new world which was, as a contemporary commentator described, replete with "antipodean perversities" (Finney 1984).

Rare conservatory plants were commonplace; the appearance of light-green meadows lured settlers into swamps where their sheep contracted rot; trees retained their leaves and shed their bark instead; the more frequent the trees, the more sterile the soil; birds did not sing; swans were black, the eagles white; bees were stingless; some mammals had pockets, others laid eggs; it was warmest on the hills and coolest in the valleys, even the blackberries [wild raspberries] were red, and to crown it all the greatest rogue may be converted into the most useful citizen: such is *Terra Australia*.

(Martin 1838, in Powell 1976: 13–14)

Martin would have undoubtedly found himself in sympathy with Major Robert Ross, Lieutenant Governor of New South Wales, who stated in 1788, "I do not scruple to pronounce that in the whole world there is not a worse country where nature is reversed" (in Hall 1992). Similarly, the French explorer Baudin was aghast at the primitive nature of the western coast of New Holland: "In the midst of these numerous islands there is not anything else to delight the mind; the aspect is altogether the most whimsical and savage; it is truly frightful" (in Marshall 1968: 9). However, the romantic image would come to predominate in Europe. Why?

The answer lies embedded in the means to which such images are put. Official images of the Pacific emphasized the romantic and the picturesque for two major reasons. First, such an image was in keeping with the dominant intellectual fashion of the times. Second, images could be put to utilitarian ends. Government utilized and encouraged such images in order to encourage settlement and therefore provide a firmer rationale for the incorporation of these new lands into the imperial structures and, ultimately, into the global system.

Australia was advertised in Britain as "a land of promise for the adventurous—a home of peace and independence for the industrious—an El Dorado and an Arcadia combined" (Sidney 1852, in Powell 1976: 21). Paralleling the ideas of Jefferson in the United States, the establishment of an idyllic rural landscape replete with yeoman farmers was a major objective of the early Australian colonial governments.

> For after all, in our Australian paradise, our highest aim is to exhibit on a small scale something like the beauties which rise up at every step in the land to which we have bade adieu, well content if we can here and there produce a corn-field surrounded by a post and rail fence, or a meadow of English grasses clear of stumps.
>
> (Curr 1824: 11–12)

The idea of a Pacific Arcadia was, therefore, developed in order to encourage a flow of migrants to the new worlds of the Pacific. So powerful was the initial promotion of this image of a better life in this world, that it continues to this day, at least in the European and North American mind and in the promise of tourist advertising. The ability of tourism to utilize such an evocative image should come as no surprise, for the promise of a better time is little different in either migration schemes or tourist packages.

Moreover, the agencies responsible for migration and the encouragement of international tourism were often one and the same. For example, the Immigration and Tourist Bureau of New South Wales was under the New South Wales Premier's Department until 1919, when the immigration function was left to the Commonwealth Government (Horne 1991). In addition, the transport system which brought the migrants and provided for international trade was the same system which served the early tourist. Thus Fiji has been a tourism destination since the early twentieth century when it was a regular stopping point for trans-Pacific shipping.

The economic potential of tourism was officially recognized in 1924 when the Fiji Publicity Board was established to run a tourist bureau at the behest of the White Settlement League. The terms of reference for the Board were "to make recommendations with a view to popularizing the colony to tourists, to provide facilities to tourists to visit places of interest, and to consider the best suitable methods of providing funds for the objects it desired to attain" (Ministry of Tourism 1992: 1). Similarly, mercantile shipping connections between Hawaii and the United States mainland served as the basis for both the annexation of the islands by the United States and the development of a tourism industry, to which commercial interests began to apply the term "Paradise" in the 1850s (Douglas and Douglas 1996).

The shipping network that overlay ancient migration and trading routes in the South Pacific was extremely important in creating a Pacific identity in the late nineteenth and early twentieth centuries. It effectively tied together the islands of the South Pacific with the metropolitan powers of Australia and New Zealand to the west, and with Canada and the United States to the east. Despite the vast distances and the different cultures and peoples in this region—Melanesia, Micronesia and Polynesia, with Australia and New Zealand at the periphery—all were drawn together into a common region in the eyes of the European and North American gaze, a regional and cultural identity that was reinforced in the production of tourism images that remain with us today.

For example, the Australian and New Zealand promotional campaign for Vanuatu in the early 1990s used the theme of "Vanuatu the untouched paradise" and featured Australian musician John Farnham's hit song "Touch of Paradise." According to the National Tourism Office of Vanuatu (1990) the campaign led to "an immediate increase in visitors from Australia which showed very healthy increases in the second half of 1989." However, as the *Pacific Islands Monthly* observed, "sceptics may smile at both the originality and the accuracy of the slogan. Surely of all the Pacific's 'paradises,' Vanuatu has been touched more often than many? But the success of the campaign is beyond argument" (1990: 38). Yet, as Douglas and Douglas argue,

> The myth of Paradise is by now a thoroughly shop-worn cliché, which invests every kind of promotion. Virtually every travel brochure on the region contains similar images, no longer the exclusive preserve of Tahiti, which inspired them, or Hawai'i which mass produced them. By the 1970s, aided by jet travel, packaged vacations and the relentlessness of brochure and television advertising, the myth had been exported more widely than any other regional product and was being applied indiscriminately and often incongruously to every part of the Pacific.
>
> (Douglas and Douglas 1996: 32–3)

Indeed, they went on to note that the "myth had become so pervasive that its presence was evident even in the work of those who ought to be critical of it" (*ibid.*: 34), and illustrated this by noting that Farrell, in his introduction to *Hawaii: The Legend That Sells*, is lured to its use thus: "Take a group of breathtakingly beautiful islands set in the blue Pacific as close to paradise as you wish" (Farrell 1982: xiii). Yet all is not well in paradise.

Globalization and the Pacific

The problems which face the island nations of the South Pacific are typical of those confronting nearly all of the world's island micro-states (Milne 1992). Because of their colonial history, small size and distance from major markets, the Pacific island nations have few viable economic resources and an extremely small indigenous capital base. As a consequence, Pacific island economies generally share one specific feature in common: "they are net importers with minimal capacity to independently generate foreign exchange" (Department of Foreign Affairs and Trade 1994). They are therefore reliant on foreign powers to provide capital for economic development and the transport links that enable the export of goods and services. In addition, most have relatively little control over their natural resources, and even less power to influence the economic and political direction of the region (Fairbairn 1985; Hall and Page 1996).

The few natural resources which can be exploited, such as fish, minerals and timber, are rapidly dwindling due to the lack of economic alternatives and development options (Fairbairn 1991; Crocombe 1992; Milne 1992; Hall 1997). Furthermore, many of the economies of the Pacific island countries are centered on one or two commodities that are typically subject to extreme price fluctuations:

> The outcome is that they generally have very few products to sell to sophisticated markets. There exists a small fragile private sector to pursue opportunities as they arise, but obtaining suitable, quality venture capital to finance sustainable globally competitive ventures is a chronic problem.
>
> (Department of Foreign Affairs and Trade 1994)

It is perhaps not surprising, therefore, that Pacific island governments, given the need to diversify their economic bases, rising social expectations and increasing population pressures, attach great importance to the development of service industries such as tourism, as a means of making an important contribution to economic growth and employment (Dorrance 1986; Department of Foreign Affairs and Trade 1994; Hall 1997). As Connell (1988: 62) stated: "For island states that have very few resources, virtually the only resources where there may be some comparative advantage in favour of [island micro-states] are clean beaches, unpolluted seas and warm weather and water, and at least vestiges of distinctive cultures." The destinations and communities of the Pacific find themselves, consequently, in a paradox.

Tourism, amongst other trade in commodities and services, helped draw the Pacific into the global capitalist system. Now that it is a part of the system, albeit at the edge of the global economy, tourism is one of the few possibilities by which much of the Pacific can hope to gain the benefits of modernity. This situation has been likened by some commentators to a situation of economic and cultural dependency or as a form of imperialism.

For example, Crick argues that tourism is a form of "leisure imperialism" and represents "the hedonistic face of neocolonialism" (1989: 322). Similarly, Nash perceived the concept quite broadly:

> At the most general level, theories of imperialism refer to the expansion of a society's interests abroad. These interests—whether economic, political, military, religious, or some other—are imposed

on or adopted by an alien society, and evolving intersocietal transactions, marked by the ebb and flow of power, are established.

(Nash 1989: 38)

Thus the importance of the relationship between the metropolitan center and the periphery was identified as follows:

Metropolitan centers have varying degrees of control over the nature of tourism and its development, but they exercise it—at least at the beginning of their relationship with tourist areas—in alien regions. It is this power over touristic and related developments abroad that makes a metropolitan center imperialistic and tourism a form of imperialism.

(*ibid.*: 39)

Imperialism may be conceived of as "the relationship of a ruling or controlling power to those under its dominion. What we mean when we speak of empire or imperialism is the relationship of a hegemonic state to peoples or nations under its control" (Lichtheim 1974: 10). However, the extent to which power is able to be exercised, and hence development controlled in any nation or destination by an external agency, is somewhat problematic as a more complex notion of globalization has replaced simplistic ideas of imperialism. Indeed, there is a general failure of critics of cultural imperialism to grasp fully the ambiguous gift of capitalist modernity inherent in contemporary globalization, that is, to probe the contradictions of capitalist culture and its implications for tourism (Britton 1991).

The idea of imperialism contains the notion of the intended spread of a social system from one center of power across the globe:

The idea of globalisation suggests interconnection and interdependency of all global areas which happens in a less purposeful way. It happens as the result of economic and cultural practices which do not, of themselves, aim at global integration, but which nonetheless produce it.

(Tomlinson 1991: 175)

Imperialism may well have been a useful description of the process by which the Pacific was incorporated into the global system itself, but a fuller understanding of the totality of cultural, economic and social change needs to be located in an understanding of the process of globalization as an inevitable outcome of modernity. Indeed, one of the paradoxes of globalization is that it implies the decay of previous imperial powers.

The idea of cultural imperialism drew on the image of relatively secure cultural communities exercising influence over other 'weaker' cultures. As national governments in late modernity are less and less able to act autonomously in the political-economic sphere, all this changes. When people find their lives more and more controlled by forces beyond the influence of those national institutions which form a perception of their specific "polity," their accompanying sense of belonging to a secure culture is eroded.

(*ibid.*: 176)

The creation of the Pacific in terms of otherness, identity and, clearly, as a destination, therefore lies in an understanding of the cultural politics of globalization. Cultural identity is "an ongoing process, politically contested and historically unfinished" (Clifford 1988: 9). Tourism is clearly inseparable from such cultural politics, which can be defined as the struggles over the official symbolic representations of reality that shall prevail in a given social order at a given time.

One could argue that these are the most important kind of politics, for they seek to "control the terms in which all other politics, and all other aspects of life in that society, will take place" (Ortner 1989: 200, in Wood 1993). Nevertheless, tourism should not be observed in isolation, as "tourism inevitably enters a dynamic context, and in the process contention over definitions of what is traditional and authentic becomes charged with a variety of additional meanings, as the range of interested parties increases" (Wood 1993: 63–4).

Tourism is very much a child of modernity. In the (post-?) modern world, tourism provides the "other" and a reference point for identities. Culture, identity and representation are continually being invented and reinvented by both insiders and outsiders. The significance of tourism "resides in the connections and disconnections it constitutes in the general processes of social change" (Hughes-Freeland 1993: 138). Tourism neither "destroys" culture nor does it ever simply "preserve" it. Instead, "[i]n a fundamental sense, both today's and tomorrow's cultural tourists seek out not pre-development culture but the outcomes of different discourses and modes of development" (Wood 1993: 68). Therefore, as Tomlinson (1991: 28) has recognized:

we will have to problematise not just those cultural practices characterized as "modern," but the underlying cultural "narrative" that sustains them: a narrative rooted in the culture of the (capitalist) West, in which the abstract notions of development or "progress" are instituted as global cultural goals.

However, the realm of modernity is problematic in terms of identity (Giddens 1981). The general sense of cultural belonging is replaced, in the "stable mode" of capitalist modernity, by a "commodified" habitual social experience in which all "identities" become effectively submerged (Tomlinson 1991). The "us" of "in-group identification" becomes increasingly difficult to fill with a content, other than those specifically invoked in the ideology of nationalism, ethnicity, religion and heritage in the increasing "sameness" of commodified modern life. Such an observation is extremely significant within the context of heritage management and heritage tourism, which often provide an idealized past while ignoring a number of competing histories (Hall and McArthur 1996; Tunbridge and Ashworth 1996). Yet this commodified representation of individual places serves as the basis for much of the promotion and development of the Pacific as a tourist destination (Hall and Page 1996; Hall 1997).

How we "live" is never a "static" set of circumstances, but always something in flux, in process. The political discourse of national culture and national identity *requires* that we imagine this process as "frozen," and this is done via concepts like the "national heritage" or our "cultural traditions." This "freezing" conceals a complex historical process in which sorting out the definitive features of "our culture" becomes highly problematic.

> What we take to be "our culture" at any time will be a kind of "totalisation" of cultural memory up to that point. This totalisation will be a particular and selective one in which political and cultural institutions (the state, the media) have a privileged role; as a consequence, "our culture" in the modern world is never purely "local produce," but always contains the traces of previous cultural borrowing or influence, which have been part of this "totalising" and have become, as it were, "naturalised."
>
> (Tomlinson 1991: 90–1)

Although there is evidence that an unprecedented cultural convergence appears to be occurring at certain levels—as does its corollary, localization, and perhaps nationalism and fundamentalism—this cannot be inferred in the transparently negative terms that the critics of "cultural homogenization" or consumerism assume. The notion of cultural homogenization is far from simple. For those in a position to view the world as a cultural totality, it cannot be denied that certain processes of cultural convergence are under way, and that these are new processes: "The world system, rather than creating massive cultural homogeneity on a global scale, is replacing one diversity with another; and the new diversity is based relatively more on interrelations and less on autonomy" (Hannerz in Clifford 1988: 17). As Tomlinson (1991: 109) recognized:

> The problem of homogenisation is likely to present itself to the Western intellectual who has a sense of the diversity and "richness" of

global culture as a particular threat. For the people involved in each discrete instance, the experience of Western capitalist culture will probably have quite different significance.

The spread of capitalism should, therefore, be seen as the spread of a distinctive "cultural dominant" in its own right.

The contextualizing structure for global processes of tourism and the Pacific as a destination, capitalism, is more than just a set of economic practices. Instead, it is the set of *central (dominant) positioning of economic practices* within the social ordering of collective existence (Tomlinson 1991). The latest expansion of capitalism through the communications and information revolution has, consequently, produced a truly "global" system, which can be seen not only in the complex networks of international finance and multinational capitalist production, but also in the spatial context of the cultural experience which it produces.

One of the great apparent paradoxes of contemporary globalization is the extent to which the local or localization has become significant. As Porter (1990: 19) noted, "Competitive advantage is created and sustained through a highly localized process. Differences in national economic structures, values, cultures, and institutions, and histories contribute profoundly to competitive success." Similarly, in the socio-political sphere, Jacques (1989: 237) observed "As power moves upwards from the nation-state towards international unity, so there is a countervailing pressure, whose roots are various, for it to move downwards." Such a situational dynamic has added a new dimension to the Pacific as destination.

Long dominated by European and North American views of the Pacific, the new global order has provided the opportunity for greater self-definition among those who reside in the Pacific. Most notably, economic interdependence has assisted in the development of the concept of the Pacific Rim, or Pacific Community, which has led to the creation of a series of new economic and political fora (Hall 1997), while improvements in communications technologies have enabled small operators in the Pacific to directly interact with their consumers, thereby bypassing wholesalers in the metropolitan regions (Milne 1996).

The reorganization of the international economic order has created new centers and peripheries, as well as new territorial hierarchies. With this spatial reorganization, Pacific tourism—and the definition of the Pacific as a destination—is entering a new era. This is not to argue that everything is and will be well with the Pacific. Rather, it is intended to re-emphasize the emergent tourism opportunities as the global system is redefined.

Cultural restructuring is as much a factor of globalization as economic restructuring. Yet long-standing myths and images surrounding the Pacific as a destination may be hard to change. In the new cultural space of (post-?) modernity "there is a new search for identity and difference in the face of impersonal global forces" (Jacques 1989: 237). Faced with the uncertainty of globalization, the advertisers' promise of a Pacific paradise with a pleasant environment and simpler life may seem more attractive to consumers than ever.

References

Agnew, J. A. (1987) *The United States in the World-Economy: A Regional Geography*, Cambridge and New York: Cambridge University Press.

Bell, D. (1979) *The Cultural Contradictions of Capitalism*, London: Heinemann.

Britton, S. G. (1991) "Tourism, capital and place: towards a critical geography of tourism," *Environment and Planning D: Society and Space*, 9 (4): 451–78.

Brown, G. H. (1988) *Visions of New Zealand: Artists in New Zealand*, Auckland: David Bateman.

Clifford, J. (1988) *The Predicament of Culture: Twentieth-Century Ethnography, Literature and Art*, Cambridge MA: Harvard University Press.

Connell, J. (1988) *Sovereignty and Survival: Island Micro states in the Third World*, Research Monograph no. 3, Sydney: Department of Geography, University of Sydney.

Crick, M. (1989) "Representations of international tourism in the social sciences: sun, sex, sights, savings, and servility," *Annual Review of Anthropology*, 18: 307–44.

Crocombe, R. (ed.) (1992) *Culture and Democracy in the South Pacific*, Suva, Fiji: Institute of Pacific Islands, University of the South Pacific.

Curr, E. (1967) [1824] *An Account of the Colony of Van Diemen's Land Principally Designed for the Use of Emigrants*, London: George Cowie; facsimile, Hobart: Platypus Productions.

Department of Foreign Affairs and Trade (1994) *Trade Patterns—South Pacific*, Canberra: Department of Foreign Affairs and Trade.

Dorrance, G. S. (1986) "The desirability of tourism," *Pacific Economic Bulletin*, 1, 2: 47–8.

Douglas, Ngaire and Douglas, Norman (1996) "Tourism in the Pacific: historical factors," in C. M. Hall and S. Page (eds) *Tourism in the Pacific: Issues and Cases*, London: International Thomson Business Press, 19–35.

Fairbairn, T. I. J. (1985) *Island Economies: Studies from the South Pacific*, Suva, Fiji: Institute of Pacific Studies of the University of the South Pacific in association with Asia Pacific Research Unit.

——(1991) *The Pacific Islands: Politics, Economics, and International Relations*, Honolulu: East-West Center, International Relations Program.

Farrell, B. (1982) *Hawaii: The Legend that Sells*, Honolulu: University of Hawaii Press.

Finney, C. M. (1984) *To Sail Beyond the Sunset: Natural History in Australia 1699–1829*, Adelaide: Rigby.

Friedman, J. (1994) *Cultural Identity and Global Process*, London: Sage.

Giddens, A. (1981) *A Contemporary Critique of Historical Materialism, Volume One: Power, Property and the State*, London: Macmillan.

Hall, C. M. (1992) *Wasteland to World Heritage: Preserving Australia's Wilderness*, Carlton VIC: Melbourne University Press.

——(1997) *Tourism in the Pacific Rim: Development, Impacts and Markets*, 2nd edn, South Melbourne: Longman Australia.

Hall, C. M. and McArthur, S. (eds) (1996) *Heritage Management in Australia and New Zealand: The Human Dimension*, Melbourne: Oxford University Press.

Hall, C. M. and Page, S. (eds) (1996) *Tourism in the Pacific: Issues and Cases*, London: International Thomson Business Press.

Hitchcock, M., King, V. T. and Parnwell, M. J. G. (1993) "Tourism in South-East Asia: introduction," in M. Hitchcock, V. T. King and M. J. G. Parnwell (eds) *Tourism in South-East Asia*, London and New York: Routledge, 1–31.

Hoogvelt, A. (1982) *The Third World in Global Development*, London: Macmillan.

Honour, H. (1981) *Romanticism*, Harmondsworth: Penguin.

Horne, J. (1991) "Travelling through the romantic landscapes of the Blue Mountains," *Australian Cultural History*, 10: 84–98.

Hughes-Freeland, F. (1993) "Packaging dreams: Javanese perceptions of tourism and performance," in M. Hitchcock, V. T. King and M. J. G. Parnwell (eds) *Tourism in South-East Asia*, London and New York: Routledge, 138–54.

Jacques, M. (1989) "Britain and Europe," in S. Hall and M. Jacques (eds) *New Times: The Changing Face of Politics in the 1990s*, London: Lawrence and Wishart.

Knox, P. and Agnew, J. (1989) *The Geography of the World-Economy*, London: Routledge.

Laurier, E. (1993) "Tackintoshary: Glasgow's supplementary glossary," in G. Kearns and C. Philo (eds) *Selling Places: The City as Cultural Capital, Past and Present*, Oxford: Pergamon Press, 267–90.

Lichtheim, G. (1974) *Imperialism*, Harmondsworth: Penguin.

Marshall, A. J. (ed.) (1968) *The Great Extermination: A Guide to Anglo-Australian Cupidity, Wickedness and Waste*, London: Panther Books.

Milne, S. (1992) "Tourism development in Niue," *Annals of Tourism Research*, 19, 3: 565.

——(1996) "Tourism marketing and consumer reservation systems in the Pacific," in C. M. Hall and S. Page (eds) *Tourism in the Pacific: Issues and Cases*, London: International Thomson Business Press, 109–29.

Ministry of Tourism (1992) *General Information on Tourism in Fiji: Its Past and Future and Impact on the Economy and Society*, Suva, Fiji: Ministry of Tourism.

Nash, D. (1989) "Tourism as a form of imperialism," in V. Smith (ed.) *Hosts and Guests: The Anthropology of Tourism*, 2nd edn, Philadelphia PA: University of Pennsylvania Press, 37–52.

National Tourism Office of Vanuatu (1990) *A History of Tourism in Vanuatu: A Platform for Future Success*, Port Vila: National Tourism Office of Vanuatu.

Ortner, S. B. (1989) "Cultural politics: religious activism and ideological transformation among 20th century Sherpas," *Dialectical Anthropology*, 14: 197–211.

Pacific Islands Monthly (1990) "Vanuatu's revival as untouched paradise," 60, 3: 37–9.

Porter, M. E. (1990) *Competitive Advantage of Nations*, New York: Free Press.

Powell, J. M. (1976) *Environmental Management in Australia, 1788–1914: Guardians, improvers and profit, an introductory survey*, Melbourne and New York: Oxford University Press.

Rousseau, J.-J. (1978) *The Social Contract and Discourses*, trans. G. D. H. Cole, rev. J. H. Brumfitt and J. C. Hall, Everyman's Library, London and New York: Dent, Dutton.

Seddon, G. (1976) "The evolution of perceptual attitudes," in G. Seddon and M. Davis (eds) *Man and Landscape in Australia: Towards an Ecological Vision*, Canberra: AGPS, 9–17.

Selwyn, T. (1993) "Peter Pan in South-East Asia: views from the brochures," in M. Hitchcock, V. T. King and M. J. G. Parnwell (eds) *Tourism in South-East Asia*, London and New York: Routledge, 117–37.

Smith, B. (1960) *European Vision and the South Pacific 1768–1850: A Study in the History of Art and Ideas*, Oxford: Clarendon Press.

Tawney, R. H. (1938) *Religion and the Rise of Capitalism*, Harmondsworth: Penguin.

Tomlinson, J. (1991) *Cultural Imperialism: A Critical Introduction*, Baltimore MD: Johns Hopkins University Press.

Tunbridge, J. E. and Ashworth, G. J. (1996) *Dissonant Heritage: The Management of the Past as a Resource in Conflict*, Chichester: Wiley.

Wood, R. E. (1993) "Tourism, culture and the sociology of development," in M. Hitchcock, V. T. King and M. J. G. Parnwell (eds) *Tourism in South-East Asia*, London and New York: Routledge, 48–70.

9

THE SOCIAL CONSTRUCTION
OF TOURIST DESTINATIONS

The process of transformation of the Saariselkä
tourism region in Finnish Lapland

Jarkko Saarinen

Constant change and transformation are typical of tourism as a spatial
phenomenon. This change has as much to do with the physical and representa-
tional base of tourist destinations as with the identity and cultural self-images
of people living in the region. The socio-spatial portrayals of reality produced
by the tourism industry can be problematic from the viewpoint of the local
inhabitants and cultures, however, as the images often express the values and
institutional practices of a non-local tourism industry. Consequently, it is
necessary to consider how a destination and its representations emerge and
how they are transformed and redefined in the context of culturally and
socially sustainable tourism.

This chapter considers the transformation of the destination as both a social
construct and a geographical process. The aim is to conceptualize the development of destinations and the institutional practices by which they are
constructed, as well as their relation to the local culture and identity. The
theoretical framework is based on the theory of the institutionalization of
regions put forward by Paasi (1986), who suggests that regions are constituted
through historically contingent processes as part of the transformation of
larger socio-spatial structures. As a region, a tourist destination represents a
specific historical and cultural phase in society and is best understood through
this temporal and social context. Using the Saariselkä tourism region in
Finnish Lapland—in the territory of the Sami, an indigenous people living in
northern Scandinavia—as a case-study, the following study analyzes the transformation of Saariselkä from a reindeer husbandry area to a roadless wilderness
and finally to a modern tourist destination. There are about 6,400 Sami in
Finland, and they possess a cultural history and identity of their own, as
revealed in the practiced forms of their livelihood.

Tourist destinations and the production of space

Tourist destinations

"Destination" is a problematic concept in practical terms, due in part to its existence at a variety of spatial scales, and as a range of possible tourism products (see Boniface and Cooper 1987; Ashworth and Voogd 1990; Hudman and Jackson 1990). While such distinctions are useful, a tourist destination as a social construct encompasses more than administrative boundaries or physiographic features (see Ryan 1991: 51; Burns and Holden 1995: 179). It is necessary, therefore, to define the notion further.

The geographical concept of region offers one basis for defining tourist destination. A region may be understood as a historical structure which is lived, experienced and represented through administrative, economic and cultural practices (Paasi 1991, 1996: 32–5). Thus tourism region—destination —is part of a larger regional structure, mediated by the awareness of the people and other actors involved, for example tourists and tourism promoters (Saarinen 1995).

Representations and production of space

The discourse of tourism in recent years has stressed the meaning of space, and the interplay between space, as place and region, and social structures, experiences and identity (Urry 1990; Shields 1991; Rojek 1993; Ashworth and Dietvorst 1995). Such a notion is particularly relevant for the conceptualization of destinations as places of tourism, for as Urry (1995: 1) contends, "[a]lmost all the major social and cultural theories bear upon the explanation of place." From the perspective of the new (or reconstructed) regional geography, it is necessary to recognize the role of space—and the institutional practices which construct it—in order to understand the social transformation of the destination. Yet as Haywood (1986) notes, little attention is given in the tourism literature to the identification of the tourist destination, as object of analysis, and its conceptual nature (Cooper 1989).

Haywood's observation concerns Butler's (1980) classic theory regarding the life-cycle of the tourism area, whereby attention is paid to the quantitative level of tourism development while simultaneously ignoring the historical and social nature of destinations. As a result, the conceptual power of destination is treated in cavalier fashion by theoretical constructs which tie it to neither space nor history, despite Harvey's (1989, 1993) observation that geography is necessarily dependent on such ideations (see Massey 1984). Indeed, the development of tourist destinations is best understood as a process of producing spaces, or what Shields (1991) describes as the concept of social spatialization, an ongoing process constructed by institutional practices and cultural discourse.

For Shields, the production and reproduction of space and place-mythology, and the changing social spatialization of tourist sites, are actions neither neutral nor passive (Lefebvre 1991). Instead, it is a spatial struggle in which the evolving development produces tourist spaces, or landscapes representing the values and needs of the non-local tourism industry rather than local interests and identities (Bruner 1991; MacCannell 1992; Hall 1994). Among the conflicts between local and non-local values are the moral (and ethical) aspect of producing the space, for space as a social construct defines what is allowed and suited to a certain region and place (Harvey 1990; Sack 1992). The effects of this localized production of space and its relationship to tourism places can be seen in the changing representations of Sami culture produced by the Finnish tourism industry, the most popular of which are the traditional ethnic dress, reindeer husbandry and shamanistic rituals.

The term Sami, which reflects the self or "inner identity" of the indigenous population and is used locally to refer to the Lapp people, is quite new in the language of the tourism industry (see Aikio *et al.* 1994; Pentikäinen 1995: 21–2). A dictionary produced for travelers in Finnish Lapland in 1976 defines Sami as "based on the Lappish term *sabme*—a name used by the Lapps for themselves. See: Lapps" (Mäkinen 1976: 130). This definition has much in common with the early travel literature, which construed the culture of the Sami in terms of the introduced values and language of explorers, visitors and the tourism industry in Lapland (Cohen 1993; Duncan 1993a).

Jean-François Regnard, regarded as one of the first travelers of any renown, visited Lapland in 1681 and his notes of the journey, published in 1731 as *Voyage de Lapponie*, along with Olaus Magnus' *Historia de Genibus Septentrionalibus* (1555) and Johan Schefferus' *Lapponia* (1673), formed the factual basis for impressions of Lapland and its inhabitants for many years thereafter. In common with later travelers, notably in the eighteenth and nineteenth centuries, Regnard (1982) described the backwardness and simplicity of the Sami culture, the curious habits of the Sami, their pagan rituals and their general untidiness. He wrote of their alcohol addiction and of the pleasantness of the Sami women (see also Suomalaisesta . . . 1885; Linné 1969; Skjöldebrand 1986; Saarenheimo 1988). Though he generalized that "all the men and women in Lapland are extremely ugly and resemble apes," Regnard managed to attribute a certain mystique and eroticism to the Sami women, and later travelers produced and reproduced this mystic picture to create a descriptive discourse. Sir Henry George Liddell even shipped two young Lapp women together with five reindeer to England in the late eighteenth century (Hirn and Markkanen 1987).

The highly romanticized version of Sami culture is clearly illustrated in Figure 9.1 in a postcard sent from Lapland in 1903. Sensually attired in a traditional Sami dress, the woman reflects this mythical discourse of Sami women and culture. In the background is a Sami hut, the *kota*, still widely used in tourism commercials today although the Sami abandoned the *kota*

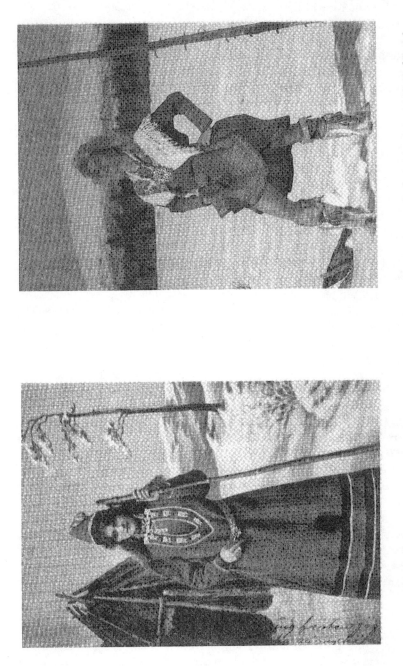

Figure 9.1 Romanticized versions of Sami culture, as illustrated in a postcard sent from Lapland in 1903 (left) and a holiday property advertisement produced at the time of tourism's rapid development in Lapland in the late 1980s (right)

several decades ago. In the adjacent panel, a holiday property advertisement produced at the time of tourism's rapid development in Lapland in the late 1980s is depicted. Again, the blonde woman posing in Sami dress alludes to a historically constructed, erotic image of Sami women, though the dress is the only culturally authentic trait depicted.

The women with their Sami dress, and the *kota* and snowy fjelds in the background, turn the picture into an ethnic landscape, or ethnoscape, that manifests the stereotyped landscape produced by the tourism industry. Examined more critically, this landscape is a product of the cultural exploitation and marginalization of Sami culture and its meaningful symbols. The blonde depicted in Figure 9.1 particularly highlights these two essential problems in the (mis)representation of Sami culture. First, the people depicted in the tourist brochures and those present at the ethnic rituals put on for tourists are not Sami but members of the dominant culture impersonating the Sami. Second, the industry is exploiting one of the main symbols of Sami culture and identity, the *gàkti* or traditional clothing, in an unsustainable manner. The *gàkti* contains information about the person wearing it as to his or her social status and residency, yet there is little consideration of these cultural meanings in the tourist replications.

Such misleading elements are problematic not only for the Sami. According to Brett (1994: 117) a culture represents a coherent picture of the inhabited world for its participants, in which they can identify their present status and past experiences. At the same time, culture is a system of self-reflections in which members become critically aware of the way in which they are represented by others (Moscovici 1988). As noted by Urry (1990: 83–6), the present situation—whether we call it postmodern or late traditional—problematizes the distinction between representations and reality (Selwyn 1996). The location of the representations is one in which there is no authenticity, and where the artificially constructed appears more real than reality itself (Duncan 1993b; Cohen 1995; Brown 1996). Certainly, this "pseudo-reality" may help to reconcile supply and demand, but misleading presentations undermine the sustainability of the destination by arousing antagonism between hosts and their visitors, eventually dictating limits of acceptability for the long term (Doxey 1975; Boissevain 1996).

The critical interpretation of representations of locality—indigenous peoples or other "host communities"—implies analyses of the actors and processes involved in the (re)production of space for the purposes of tourism (see Meethan 1996). This necessitates a historical perspective of the destination.

Transformation of the destination

The theory of the institutionalization of regions (Paasi 1986) constitutes a fruitful framework for analyzing the construction of tourist destinations as elements of socio-spatial reality. The theory of institutionalization describes how regions and their boundaries—physical or mental—emerge, are transformed and disappear in the course of the regional transformation. Paasi proposes a conceptual distinction between region and place, whereby place is defined individually through personal interpretations of spatial phenomena, and by the actions of the individual (see Johnston 1991). Region, however, is interpreted collectively and represents institutional practices and the history of the region rather than the history of the individual and place. Thus the difference between place and region is not one of geographical scale but of historical construction (Gilbert 1960).

The emergence of a region may be conceptualized in four stages which illustrate the process of transformation (Paasi 1986): territorial shape, symbolic shape, institutional shape and established role. In the transformation of a tourist destination, there is little need to strictly distinguish between territorial and symbolic shapes, however, as destinations (as defined here) are distinguished from their surroundings and from other destinations more or less conceptually, without exact physical or territorial boundaries. The territorial and symbolic stages thus merge to form a "discourse of region," a process in which the socio-spatial meanings and representations characterizing the destination are produced and reproduced. The main factors comprising the discourse include travelogues and regional literature, tourist advertisements and the media in general. Along with the tourists themselves, these elements make the natural and cultural features of the destination known. At the same time, the process stereotypes and modifies the socio-spatial representations (Cohen 1993; Dann 1996).

The third element of Paasi's theory, institutional shape, reflects the political and economic processes that produce a discourse of region. A region is linked by its institutional shape to a larger regional and economic structure, and finally to the world economy and politics, as inseparable parts of these. Locally, these processes are manifested in the context of the various organizations set up for tourism planning, development and marketing, in specific touristic enterprises, and in the consumption of goods and services. Since the goal of the participants is to promote tourism, this stage may be termed the "discourse of development," characterized by increased numbers of tourists, improvements to the infrastructure and the stimulation of consumption. While traditional research methodologies, for example Butler's evolution cycle (1980), may prove useful in describing the development, the discourse of development does not presume that each stage necessarily follows a preordained order. Instead, the transformation of a tourism destination may be seen as part of a larger social and economic structure in which the nature of change is not easily assumed.

The achievement of an established role, the fourth component, refers to the stage of institutionalization at which a region has acquired an identity which comprises a material basis and socially constructed representations. The iconographic meaning and history of the destination are continually produced anew in order to attract tourists and to distinguish the region from other destinations. This established role is not, in one interpretation, an independent discursive structure, but the result of an encounter between the discourses of region and development, an abstraction reflecting the present identity of the destination. Like Williams' (1988) idea of hegemonic culture, this identity contains features from the present, traces from the past and signs of future development. Thus it becomes important to be able to describe and analyze the processes through which the destination and its representations are constructed, named, framed and elevated, enshrined and reproduced mechanically and socially (see MacCannell 1976: 44–5).

In contrast to MacCannell's theory of sight sacralization, the destination is not a stable, permanent socio-spatial structure, but a cultural landscape subject to continual transformation and reformation, in which it emerges, changes, disappears and re-emerges in varied forms. In this manner, destinations are institutionalized and de-institutionalized, reproducing social structures and meanings through the adoption and adaptation of previous ones. Clearly then, the discursive practices through which the development of a tourist destination is constructed are neither irrelevant nor independent (see Barnes and Duncan 1992: 3–4). Instead, the distinctions between the discourse of region and those of development are conceptually powerful and socially significant to the understanding of destinations.

Case-study: the process of transformation of the Saariselkä tourism region

> I want to escape all this rushing, time and space. To a place, where I can find solitude and social opportunities as I choose. Somewhere, where facilities and solitude are both equally available.
>
> (tourist brochure, Saariselkä, 1992)

The Saariselkä tourism region is located in northeastern Lapland, alongside the major travel route to the North Cape (Figure 9.2). The region is owned by the Finnish government and administered by the Forest Research Institute and the Forest Service. The primary recreational district is the Saariselkä resort, where the main facilities and services are situated. Other tourist sites include the outdoor centers of Kiilopää and Kakslauttanen, and the Laanila and Tankavaara areas.

Saariselkä and the surrounding region now attract an estimated 200,000 tourists a year, with the current accommodation capacity at approximately 7,000 persons. Most of these facilities are owned and managed by trade unions, companies and private individuals. The annual revenue in the region from tourism was FIM109 million (US$21 million) in 1994, and the industry employs nearly 170 local residents from nearby districts (Saarinen et al. 1996).

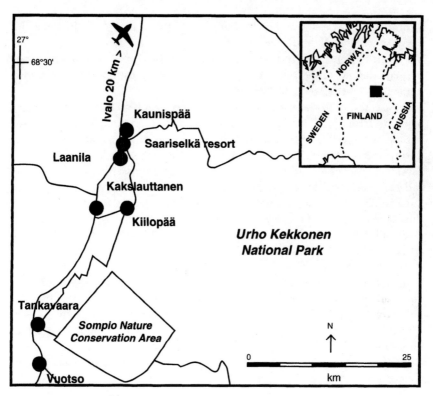

Figure 9.2 Saariselkä tourism region, northeastern Lapland

The discourse of the region

The remote Saariselkä region did not attract the attention of early travelers until the rise of the Finnish national romantic movement in the nineteenth century, though it was visited extensively by scientific expeditions anxious to explore the cultural features of the local population and related natural phenomena (Suomalaisesta . . . 1885: 148–9; Lönnrot 1980). The origin of the term Saariselkä is not known, though it is mentioned by Rosberg in his book *Lappi* in 1911 (Itkonen 1991). Physiographically, the name refers to the watershed lying east of the present tourist region known as Kaunispää or Laanila until recent times, although the region is referred to as Suolaselkä in the Finnish geography textbook of Ignatius (1890: 137). Both terms—Suolaselkä and Saariselkä—originate from the Sami name *Suoloçielgi* which means "open sea with an island," as the fjelds, when viewed from a distance, appear to rise from their surroundings like islands from the sea (Partanen 1992).

Gold was found at the site of the current Saariselkä resort in 1871, and large-scale mining emerged shortly thereafter. Laanila gained an early reputation as the Finnish Klondike, but the expectations were never realized and mining soon came to an end (Rosberg 1911; Partanen 1977). In 1912 the main office building of the mining company was converted to an inn (Partanen 1992). Though this was the first visitor facility in the area, modern tourism did not appear until the incorporation of the Petsamo area into Finland under the Treaty of Tartu in 1920.

As a gateway to the Arctic Ocean, Petsamo soon became one of the most popular attractions for both Finnish and foreign tourists (Hirn and Markkanen 1987), and the Saariselkä region became a transit area for tourists during the interwar years, thereby establishing the borders of the "real" Lapland and the Sami territory for visitors. Featuring scenic fjeld landscapes and easily accessed by automobile, the hill of Kaunispää became one of the most famous tourist sights, and views from the top were widely publicized in the travel literature of the day.

Kaunispää and the fjeld landscapes also gained fame for their reputed therapeutic powers in healing physical and mental conditions (see Suomen matkailijayhdistys 1939; Partanen 1992). Although Kaunispää was the undisputed destination and symbol of the region by the 1920s, a coherent regional image of Saariselkä—as delineated today—did not exist. Instead, only isolated localities, such as Kaunispää and Laanila, were known, each presenting its own history and meanings (Lampio and Hannikainen 1921; Lappi 1959).

Following Finnish independence in 1917, however, tourism began to gain support as a means of introducing the new state to its people and to the international community. While Finland was promoted as a country rich in natural features, cultural differences were not equally stressed and this, together with the evolving, materialistic concept of culture which prevailed at that time, created a sense of the region as an uninhabited wilderness or "no man's land," an attitude influenced strongly by the colonialistic state policy of the day (Karmo 1932: 5; Aikio 1994). Though the first postwar facilities for tourists were named after Kaunispää, it soon became clear that tourism could not depend solely on the attractions based on the Petsamo transit traffic. Consequently, in addition to Kaunispää and Laanila, the growing tourist trade began to focus on the distant Saariselkä fjelds.

One of the most prominent figures to proclaim the fjeld's attractions was the novelist Kullervo Kemppinen, particularly in his book *Lumikuru* (1958), where the fjeld and wilderness landscapes are depicted as paradise. Indeed, so celebrated were the illustrations at the time that they were turned into popular postcards, and the book had an immediate impact on visitations to the region. While the average annual increase in the 1950s was approximately 38 percent, the area experienced a 100 percent increase in tourism in the year following *Lumikuru*'s publication (Saastamoinen 1982: 49).

Kemppinen's books described the natural features of the region and its tourists in almost noble terms. The literary landscape he created, together with the increase in tourism and the nature conservation debate of the 1960s, presented Saariselkä as a true wilderness area and one of the last of its kind in Europe (Häyrynen 1979, 1989). At the same time the domestic tourism industry sought to utilize Saariselkä as a destination, and the favorable image of its fjelds enhanced the attractiveness of the Kaunispää-Laanila region. But the growth of tourism was stymied by a lack of direct access. Thus the region could be reached only by the most enthusiastic tourists, a number deemed too small and irrelevant by the industry.

To encourage greater numbers of visitors, a road was soon planned, using the US national park system as a model. However, a less expensive solution soon surfaced, in which the perceived distance of the destination was erased by applying the term "Saariselkä" to the Kaunispää-Laanila area as well. As part of the same process, Saariselkä itself was redesignated the Urho Kekkonen National Park, a change immediately institutionalized in road maps and tourist brochures. While the (mis)use of Saariselkä met with resistance at first from other organizations located in the area, the process of mechanical and social reproduction was already underway and the term Saariselkä began to represent a larger destination (see Ansamaa 1992). As Figure 9.3 illustrates, whereas Kemppinen's (1961) map of *Poropolku kutsuu* situates Saariselkä in the midst of a roadless wilderness, another map, taken from a hiking guide published by Kai-Veikko Vuoristo (1983) nearly twenty years later, shows it located adjacent to the main road and surrounded by new communities and development.

In sum, "Saariselkä" now refers to a commercially constructed destination, rather than the historic place of the local people. As such, the process of transformation differs from that presupposed by MacCannell (1976), in that the tourist region and its attractions are not permanent structures but social constructs, continually transformed through local production and reproduction. Massive institutional support has facilitated the change from wilderness to a modern tourist destination, creating a landscape of pleasure and recreation in advertisements for the region.

Manifest in the evolving tourism landscape of the 1970s and 1980s, and in the photos of travelers hiking or skiing amid nature, was the introduction of non-local actors into the local scene. As a result, the cultural connectivity between the local population and the physical landscape of Saariselkä faded even further, and early narratives of the indigenous place were soon replaced by descriptive metaphors of the changing social location: "Saariselkä— Friends All Year Round" and "Saariselkä—Between Two Wildernesses."

With the growing number of tourists the fragmentation of Saariselkä increased, and by 1990 the growth in accommodation capacity and the fierce competition between tourist destinations led to even more distinct seasonal specialization. In the early 1980s there were three distinct tourist seasons: late winter (spring), summer and autumn. But new seasons, such as the arrival of the first snow, midwinter, Christmas, and a number of social events, were established; and hotels, restaurants and a fashionable spa and recreation center

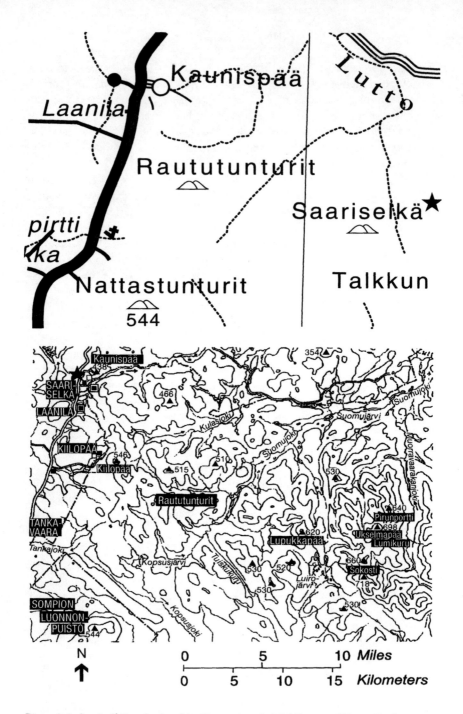

Figure 9.3 Saariselkä as depicted in Kemppinen's (1961) map of *Poropolku kutsuu* (top image), in which it is situated in the middle of a roadless wilderness; and a subsequent map published by Kai-Veikko Vuoristo (1983) nearly twenty years later, in which it is located adjacent to the main road and surrounded by new communities and development (bottom image)

were built during the late 1980s, and Saariselkä soon evolved into a more modernistic tourism landscape. Consequently the region became, during the main tourist seasons and at weekends, the *topos* of a set of connected signs and discourses of pleasure in a wilderness context.

One example of this *topos* and the combining of old and new attractions is exemplified in Figure 9.4, in which a scantily clad woman gazes at the wild landscape with its hill and ponds in their autumn colors. The wilderness, and admiration of the aesthetic beauty of Saariselkä, provide a continuum of the tourist history of the region, from the view from Kaunispää in the 1920s and 1930s and the wilderness landscape of Kemppinen's postwar books and post-cards, to the woman as product of tourism's development and the perceptual change of Lapland that occurred in the late 1980s. As such, this woman's portrayal in the tourist brochure accurately reflects the transformations affecting Saariselkä and its present identity, a process that has territorialized it as a recreational destination linked to an established conservation area—a wild setting deliberately maintained and perpetuated, albeit artificially, by erecting dead trees between the hotels to symbolize untouched nature within the resort infrastructure.

The discourse of development

The Second World War had a major impact on the future of tourism in Saariselkä. As a result of that war, Petsamo was lost to the USSR and the phys-ical infrastructure was totally destroyed. Consequently, the region emerged as a destination in its own right, rather than a transit stop on the way to Petsamo. The first postwar tourist facility was the Alamaja cottage located south of Kaunispää, dating from 1949 and owned by the Finnish Travel Association. Another cottage, Ylämaja, was built further upland by the Lapland Travel Association in 1952 (Partanen 1984).

Even more important than the first facilities in stimulating Saariselkä's development as a destination was the health situation in Lapland, where the most serious problem was tuberculosis among children. In view of the urgent needs, the Friends of Children association was founded in Lapland in 1948, and immediately began to organize childrens' summer camps in healthier environments. Kaunispää, the present Saariselkä resort, was chosen as a loca-tion for one such camp. The first facilities and the main building were ready by 1953, and initially lodged forty persons, a capacity later increased in 1957 to sixty persons. At the end of each summer season, the facilities were then made available for use by tourists (Koskela 1990).

It was no coincidence that the summer camps were held at Kaunispää. A scheme had already been launched in the 1920s to build a medicinal spa in the area because of the reputed healing benefits of local springs, but the project fell through (Partanen 1992). Thus when the association began looking for a place that typified natural health, it sought a position with elevation and wide

RUSKARAKKAUTTA

Figure 9.4 Combination of old and new attractions, in which a woman gazes at the wild landscape with its hill and ponds in their autumn colors

areas of pine forest. Despite its distance (250km) from Rovaniemi, the capital of the province of Lapland, Kaunispää's physiographic landscape—and, naturally, its location beside the main road—satisfied the criteria.

The buildings owned by the Friends of Children association were destroyed by fire in 1958. As they had been duly insured, it would have been possible to build new facilities, but the improvement in medical care and the standard of living during the 1950s deprived the camps of their medicinal significance. The association, therefore, decided to use the insurance money to build the infrastructure for commercial tourism, and a new building with modern facilities and accommodation for fifty-eight persons was opened to the public in 1960. This was the final starting-point for modern tourism in the Kaunispää-Laanila area, and saw the arrival of more tourist businesses and facilities.

Tourism in the region developed rapidly during the 1960s. Viewed in a wider context, this development was related to the growth in the world and national economy, while at a more local level, trade unions and companies in southern Finland heavily promoted construction of Kaunispää-Laanila as a tourist attraction in the 1960s and later. Certainly, the increase in leisure time and disposable capital associated with the economic development attracted trade-union members and created a small-scale industry in facilities owned by their unions and companies (Luoma 1992; Palomäki 1992).

The growing demand for building lots and the growing conflicts between nature conservation, tourism and forestry created a need for long-term land use planning and the establishment of planning committees that included representatives of many government organizations, trade and employers' unions, companies and the tourist industry (Partanen 1992). A land use plan for the area was completed in 1972, with the main goal of channeling future development into a more compact resort structure and increasing accommodation capacity, a rather new and atypical goal for a tourist infrastructure in Lapland. Based on the experiences of ski resorts in the Swiss Alps, where large, compact resorts have proven more economical and sustainable, the plan set the potential accommodation capacity of Saariselkä at approximately 8,000 beds in an area of less than two square kilometers (Maanmittauslaitos 1972).

Along with the development of tourism, the movement to conserve the remaining wilderness areas of Saariselkä grew more intense, and several proposals to this end were made by the Finnish Association for Nature Conservation and other environmental and recreational organizations during the 1960s. Their efforts finally succeeded in establishing the area as the Urho Kekkonen National Park in 1983. There was strong resistance from the local authorities, however, for whom conservation represented a continuation of non-local domination and the historically constructed colonialism exercised by southerners.

The oil crisis in the mid- and late 1970s slowed down the growth of tourism worldwide, and its effects in the Saariselkä region were equally felt, though the region subsequently recovered in the early 1980s. By the end of the decade, Finland was undergoing a period of rapid economic development, and the opening of new financial markets in the mid-1980s was reflected in the sharp increase in construction work. Both construction and the growth in tourism declined rapidly from 1991–5, however, when the deepest recession in Finland in postwar times commenced.

While the number of Finnish tourists to the region has declined, the number of foreign tourists has increased, due in part to the collapse of the German Democratic Republic as well as the devaluation of the Finnish mark. Already, construction of new hotels is underway, and the present land use plan increases lodging capacity from the current level of 7,000 beds to 22,000 beds.

Saariselkä and local representations

The transformation process and the representations of Saariselkä as a modern tourist destination reflect a landscape of tourism and hegemonic discourse that is of non-local making, and controlled by non-local actors, the influence of external capital, and global trends and patterns of travel and consumption. The developments already taking place, and the future visions of the tourist industry—whether realistic or fantasy—have had some impact on the local population and on peoples' identity and sense of place. There remain local discourses markedly at odds with the hegemonic ones and with the present identity of Saariselkä, including conflicts concerning the relationship between tourism and local culture. While fundamentally a matter affecting the entire tourist industry of Finnish Lapland rather than a particular destination, it has its local manifestations.

During the 1996 Christmas season, the Association of Young Sami distributed handouts to foreign tourists at the Arctic Circle bearing headlines such as "Fake—fake—fake," "Indigenous peoples protest," "Stereotyped people," and "Let's protect the Sami culture." This demonstration may be interpreted as an expression of the antagonism existing between the Sami population and the tourist industry as a consequence of the culturally unsustainable manner in which Lapland is presented as a tourist destination. The fact that the conflict has not presented itself as overtly as this before now in Lapland or in the Saariselkä region may be attributed to the already pronounced marginalization of the local culture and population with respect to planning processes and spaces set aside for tourism.

The Saariselkä tourist destination lies inside the Sami territory, where they have resided for centuries, but the relationship between tourism and the Sami during the transformation process has been one of separation and alienation. The local Sami population lives well away from the resort itself and the grazing grounds of the reindeer lie beyond the area devoted to tourist activities, leading the novelist Erno Paasilinna (1985: 39) to consider Saariselkä "a tourist reservation," with little or no connection to the locality in which the Sami people reside. From the local point of view, Saariselkä is a tourism landscape converted to consumption and exploitation, from which the Sami have gained little thus far (see Helle and Särkelä 1993).

For visitors to Saariselkä, representations of Sami culture are constructed by the tourist industry, and advertising for the destination contains few elements of Sami culture other than the main "ethnic products" displayed on postcards and tourist brochures. Such cultural icons, depicting people in Sami dress posing or acting in stereotypical situations such as herding reindeer, fishing or gutting fish, may be considered harmless, provided that they are interpreted from the same viewpoint from which they are produced.

In the eyes of the local Sami people, however, these representations are at best trite and banal. The traditional Sami costume is more or less equivalent in

168

its usage to the white tie and tails of Western societies, so the idea of wearing it to herd reindeer or gut fish is ridiculous. In addition, as noted earlier, the Sami dress itself is by no means presented in a consistently authentic manner. And if the person posing in the picture is not a Sami at all, the whole thing becomes an exploitation of Sami culture rather than a sympathetic, sustainable representation of local customs.

Conclusion

The purpose of this chapter has been to discuss, both theoretically and empirically, the nature of tourist destinations and their transformation process, based on the notion that destinations are social constructs and have to be understood as such, and that destinations are subject to a constant process of transformation. The framework—the process of transformation of a tourism region—stresses the place of hegemonic discourses. Furthermore, there is not just one discourse of region or development, but several competing and even conflicting ones. This is most especially the case when local needs conflict with the development strategies of the tourist industry. Hegemonic discourse, often reflecting a non-locally constructed and controlled transformation process, may constitute marginalization of the local communities and their culture, through which signs and status of local culture in the representations of the destination are dispersed.

In the discourse and practices of development, this is most clearly manifested in the matter of rights over the land, in the processes of land use planning, in the employee–employer structure, in privatization, and in similar actions taken to provide services for tourists and to cater exclusively to their needs (Shaw and Williams 1994). Where indigenous people are concerned, however, the process may take a different direction and lead to "socialization," in which local privileges concerning the use of the land and natural resources (e.g. hunting and fishing) are commercialized by the tourist industry.

With relation to the discourse of region, marginalization leads to an absence of local culture in tourist representations, and reconstruction of the history and present identity of the destination through the history of tourism and other non-local interests (Greenwood 1989; Pratt 1992). As illustrated in the present case-study, the results of this marginalization leave little or no historical material with which to analyze or describe the transformation process and the representations of destination from a local viewpoint.

Second, marginalization may take a more visible form, through commercialization of the signs and "products" of the local culture in the form of souvenirs, postcards, brochures and advertisements. These objects are not necessarily products of real or functional cultures or communities but are imaginary and grounded in myth. As is common in the tourist industry, space and time are produced and reproduced through stretching and mixing. This

169

allows it to commercialize, market and represent local cultures and their products on a larger scale than would be possible within the scope of an original, "authentic" destination (e.g. using the Sami culture to sell Finland, and even Helsinki).

The issue of the transformation and representation of destinations is crucial from the viewpoint of locality—local people and their culture—especially where there is a danger that the relationship between tourism and local culture might be based on what Said (1978) describes as an "uneven exchange." For planning, the point of theorizing the process of transformation of a destination is to help discuss and analyze the social processes and institutional practices by which the present identity and representations of destinations are produced, reproduced and controlled. In the context of the development of socially and culturally sustainable tourism, the goal is to expose the structures of the social and physical construction of a destination and its representations, and the contradictions that may exist between these and local interests and values.

References

Aikio, P. (1994) "Mikä ja mitä on periferia?" in L. Heininen (ed.) Pohjoinen Suomen politiikassa, *Lapin yliopiston hallintoviraston julkaisuja*, 27: 3–13.

Aikio, S., Aikio-Puoskari, U. and Helander, J. (1994) *The Sami Culture in Finland*, Helsinki: Lapin Sivistysseura.

Ansamaa, O. (1992) "Työmarkkinajärjestö matkailuyrittäjäksi," in S. J. Partanen (ed) *Saariselkä*, Helsinki: Suomen Matkailuliitto.

Ashworth, G. and Dietvorst, A. (eds) (1995) *Tourism and Spatial Transformations*, Oxford: CAB International.

Ashworth, G. and Voogd, H. (1990) "Can places be sold for tourism?" in G. Ashworth and B. Goodall (eds) *Marketing Tourism Places*, London and New York: Routledge.

Barnes, T. J. and Duncan, J. S. (1992) "Introduction: writing worlds," in T. J. Barnes and J. S. Duncan (eds) *Writing Worlds: Discourse, Text and Metaphor in the Representation of Landscape*, London and New York: Routledge.

Boissevain, J. (ed.) (1996) *Coping with Tourists: European Reactions to Mass Tourism*, Providence RI: Berghahn Books.

Boniface, B. G. and Cooper, C. (1987) *The Geography of Travel and Tourism*, Oxford: Butterworth-Heinemann.

Brett, D. (1994) "The representation of culture," in U. Kockel (ed.) *Culture, Tourism and Development: The Case of Ireland*, Liverpool: Liverpool University Press.

Brown, D. (1996) "Genuine fakes," in T. Selwyn (ed.) *The Tourist Image: Myths and Myth Making in Tourism*, Chichester: Wiley.

Bruner, E. M. (1991) "Transformation of self in tourism," *Annals of Tourism Research*, 18: 238–50.

Burns, P. M. and Holden, A. (1995) *Tourism: A New Perspective*, London: Prentice Hall.

Butler, R. (1980) "The concepts of a tourist area cycle of evolution: implications for management of resources," *Canadian Geographer*, 24, 1: 5–12.

Cohen, E. (1993) "The study of touristic images of native people: mitigating the stereotype for a stereotype," in D. G. Pearce and R. W. Butler (eds) *Tourism Research: Critiques and Challenges*, London and New York: Routledge.

——(1995) "Contemporary tourism: trends and challenges," in R. Butler and D. Pearce (eds) *Change in Tourism: People, Places, Processes*, London and New York: Routledge.

Cooper, C. (1989) "Tourist product cycle," in S. Witt and L. Moutinho (eds) *Tourism Marketing and Management Handbook*, London: Prentice Hall.

Dann, G. (1996) "The people of tourist brochures," in T. Selwyn (ed.) *The Tourist Image: Myths and Myth Making in Tourism*, Chichester: Wiley.

Doxey, G. V. (1975) "A causation theory of visitors-resident irritants: methodology and research infernes," in *The Impact of Tourism*, Sixth Annual Conference Proceedings of the Travel Research Association, San Diego CA.

Duncan, J. S. (1993a) "Landscapes of the self/landscapes of the other(s): cultural geography 1991–2," *Progress in Human Geography*, 17, 3: 367–77.

——(1993b) "Sites of representations: place, time and the discourse of the other," in J. Duncan and D. Ley (eds) *Place/Culture/Representation*, London and New York: Routledge.

Gilbert, E. W. (1960) "The idea of the region," *Geography*, 45, 157–75.

Greenwood, D. D. (1989) "Culture by the pound: an anthropological perspective on tourism as cultural commoditization," in V. Smith (ed.) *Host and Guest: The Anthropology of Tourism*, 2nd edn, Philadelphia PA: University of Pennsylvania Press.

Hall, C. M. (1994) *Tourism and Politics*, Chichester: Wiley.

Harvey, D. (1989) *The Condition of Postmodernity*, Oxford: Blackwell.

——(1990) "Between space and time: reflections on the geographical imagination," *Annals of the Association of American Geographers*, 80, 3: 418–34.

——(1993) "From space to place and back again: reflections on the condition of post-modernity," in J. Bird, B. Curtis, T. Putnam, G. Robertson and L. Tickner (eds) *Mapping the Futures: Local Cultures, Global Change*, London and New York: Routledge.

Häyrynen, U. (1979) *Salomaa*, Helsinki: Kirjayhtymä.

——(1989) *Koilliskaira: Urho Kekkosen Kansallispuisto*, Helsinki: Otava.

Haywood, K. M. (1986) "Can the tourist-area life cycle be made operational?" *Tourism Management*, 7, 3: 154–67.

Helle, T. and Särkelä, M. (1993) "The effects of outdoor recreation on range use by semi-domesticated reindeer," *Scandinavian Journal of Forest Research*, 8: 123–33.

Hirn, S. and Markkanen, E. (eds) (1987) *Tuhansien järvien maa: Suomen matkailun historia*, Helsinki: Matkailun Edistämiskeskus and Suomen Matkailuliitto.

Hudman, L. E. and Jackson, R. H. (1990) *Geography of Travel and Tourism*, New York: Delmar.

Ignatius, K. E. F. (1890) *Suomen maantiede kansalaisille*, Helsinki: G. W. Edlund.

Itkonen, T. I. (1991) *Lapin Matkani*, Porvoo: WSOY.

Johnston, R. J. (1991) *Questions of Place: Exploring the Practice of Human Geography*, Oxford: Blackwell.

Karmo, B. (1932) *Petsamo kutsuu: Matkailijan opas*, Porvoo: WSOY.

Kaunispään maja (1949) *Suomen Matkailu*, 2: 15.

Kemppinen, K. (1958) *Lumikuru*, Porvoo: WSOY.

——(1961) *Poropolku kutsuu*, Porvoo: WSOY.

Koskela, N. (1990) "Tunturimaja Piispan mutkan maastoon," *Freedays*, 1: 47.

Lampio, E. and Hannikainen, L. (1921) *Petsamon opas*, Helsinki: Otava.

Lappi: opaskirja Lapissa retkeileville (1959) *Lapin läänin matkailulautakunnan julkaisuja*, 6.

Lefebvre, H. (1991) *The Production of Space*, Oxford: Blackwell.

Linné, C. von (1969) *Lapinmatka 1732* (orig. Lapplandresa, 1889) Hämeenlinna: Karisto.

Lönnrot, E. (1980) *Matkat 1823–1844*, Espoo: Weilin-Göös.

Luoma, K. K. (1992) "Tonttien kova kysyntä synnytti lomakaupungin," in S. J. Partanen (ed.) *Saariselkä*, Helsinki: Suomen Matkailuliitto.

Maanmittauslaitos (1972) Maankäytön yleissuunnitelma: Saariselän matkailukeskus, Helsinki.

MacCannell, D. (1976) *The Tourist: A New Theory of the Leisure Class*, New York: Schocken.

——(1992) *Empty Meeting Ground: The Tourist Papers*, London and New York: Routledge.

Mäkinen, V. (1976) *Lapin matkailutieto*, Helsinki: Suomen Matkailuliitto.

Massey, D. (1984) *Spatial Divisions of Labour: Social Structures and Geography of Production*, London: Macmillan.

Meethan, K. (1996) "Place, image and power: Brighton as a resort," in T. Selwyn (ed.) *The Tourist Image: Myths and Myth Making in Tourism*, Chichester: Wiley.

Moscovici, S. (1988) "Notes toward a description of social representations," *European Journal of Social Psychology*, 18, 211–50.

Paasi, A. (1986) "The institutionalization of regions: a theoretical framework for understanding the emergence of regions and constitution of regional identity," *Fennia*, 164, 1: 105–46.

——(1991) "Deconstructing regions: notes on the scales of spatial life," *Environment and Planning A*, 23, 239–56.

——(1996) *Territories, Boundaries and Consciousness: The Changing Geographies of the Finnish-Russian Border*, Chichester: Wiley.

Paasilinna, E. (1985) "Saariselän kaupungista Inarin kyläraitille," *Suomen Kuvalehti*, 43, 39–44.

Palomäki, P. (1992) "Valaistun ladun tarinat," in S. J. Partanen (ed.) *Saariselkä*, Suomen Matkailuliitto: Helsinki.

Partanen, S. J. (1977) *Kultamaiden retkeilyopas*, Helsinki: Suomen Matkailuliitto.

Partanen, S. J. (ed.) (1984) *Kaunispää-Saariselkä*, Helsinki: Suomen Matkailuliitto.

——(1992) *Saariselkä*, Suomen Matkailuliitto, Helsinki.

Pentikäinen, J. (1995) *Saamelaiset: pohjoisen kansan mytologia*, Helsinki: Suomen Kirjallisuuden Seura.

Pratt, M. L. (1992) *Imperial Eyes: Travel Writing and Transculturation*, London and New York: Routledge.

Regnard, J. F. (1982) *Retki Lappiin* (orig. Voyage de Lapponie, 1731) Helsinki: Otava.

Rojek, C. (1993) *Ways of Escape: Modern Transformations in Leisure and Travel*, London: Macmillan.

Rosberg, J. E. (1911) *Lappi*, Helsinki: Kansanvalistuseura.

Ryan, C. (1991) *Recreational Tourism: A Social Science Perspective*, London and New York: Routledge.

Saarenheimo, E. (1988) *Retki Euroopan ääreen: Giuseppe Acerbi ja hänen Lapin-matkansa 1799*, Helsinki: Otava.

Saarinen, J. (1995) "Matkailualueen hahmottuminen: kaksi näkökulmaa matkailualueen kehittymiseen" (abstract: The emergence of the tourism region: two approaches to the development of tourism regions) *Terra*, 107, 4: 189–97.

Saarinen, J., Keränen, A. and Sepponen, P. (1996) "Luonnon vetovoimaisuuteen perustuvan matkailun taloudelliset vaikutukset paikallistasolla: esimerkkinä Saariselän matkailu" (abstract: The economic effects at the local level of tourism based on the attractiveness of nature: tourism at Saariselkä as an example), in J. Saarinen and J. Järviluoma (eds) Luonto virkistys—ja matkailuympäristönä, *Metsäntutkimuslaitoksen tiedonantoja*, 619: 79–92.

Saariselkä Resort (1992) *Tourist Brochure*.

Saastamoinen, O. (1982) "Economics of the multiple-use forestry in the Saariselkä fell area," *Communicationes Instituti Forestalis Fenniea*, 104.

Sack, R. D. (1992) *Place, Modernity and Consumer's World*, Baltimore MD: Johns Hopkins University Press.

Said, E. (1978) *Orientalism*, Harmondsworth: Penguin.

Selwyn, T. (1996) "Introduction," in T. Selwyn (ed.) *The Tourist Image: Myths and Myth Making in Tourism*, Chichester: Wiley.

Shaw, G. and Williams, A. M. (1994) *Critical Issues in Tourism: A Geographical Perspective*, Oxford: Blackwell.

Shields, R. (1991) *Places on the Margin: Alternative Geographies of Modernity*, London and New York: Routledge.

Skjöldebrand, A. F. (1986) *Piirustusmatka Suomen halki Nodkapille 1799* (orig. *Voyage Pittoresque au Cap Nord, 1801–1802*) Porvoo: WSOY.

Suomalaisesta Tutkimusretkestä Sodankylään ja Kultalaan vuosina 1882–83 ja 1883–84 ynnä kuvaelmia Lapista (1885) Helsinki: J. C. Frenckell ja Poika.

Suomen matkailijayhdistys (1939) *Lapin opas matkailijoita ja retkeilijöitä varten*, Helsinki: Suomen matkailijayhdistys.

Urry, J. (1990) *The Tourist Gaze: Leisure and Travel in Contemporary Societies*, London: Sage.

——(1995) *Consuming Places*, London and New York: Routledge.

Vuoristo, K. V. (1983) *Hankien kimmallus*, Otava: Helsinki.

Williams, R. (1988) *Marxismi, kulttuuri ja kirjallisuus* (orig. Marxism and Literature, 1977) Jyväskylä: Vastapaino.

INDEX

access problems: cyberspace 40–1;
  Internet 113–14; Scotland 126, 129,
  130–4
accumulation of commodities 38–9
Adams, P. 46(n2)
advertisements 26–7; images of place
  102–3; layout rules 26–7; readership
  26; representations of women 90–1,
  156–8, 166
aesthetic management of places 21–2
Akroun, M. 68
Alford, Ethelwyn 90
Anderson, K. 6
Arcades project 36–7
architecture: postmodern 39–40;
  private/public space 39
Arctic: idea of north 105–7; as tourist
  commodity 107–9; tourist
  perceptions 109–11
"Arctic Dreams" (Lopez) 106–7
Australia, visitors' perceptions 142–4
authenticity 4, 8, 158

Bachelard, G. 41
Baedeker, Karl 25
Baffin Handbook (Hamilton) 109
Bagiarta, I. N. 67–8, 74
Bali: beach 59; cosmology 54–5; culture
  53–6, 73, 74–5; guides 63, 65,
  68–73; landscape 52, 53, 66;
  modernization 61; national symbols
  60; planning 59–60; pollution 58–9;
  religion 53, 60–1, 66; social status
  67–8; tourism 55–60, 66–7, 75–6;
  training in tourism 75–6, 77(n2);
  village life 54, 58; wood carving 58
Bali Government Tourism Office 69

Bali Tourism Study 56, 57
Balmoral 127, 131
Baudrillard, Jean 23
beaches 22, 59, 70–1
Beamish, North of England Open Air
  Museum 43–4
Bell, B. 25
Bella, L. 87
Benjamin, W. 36–40, 46–7(n3)
Between Two Worlds (Greenwald) 105–6
billiard ball model 1, 5
Blanton, D. 75
Bodenhorn, B. 8
Boniface, P. 21
Boswell, James 124
Bougainville, Louis Antoine de 142
Bowes, Surgeon 143
BPLP 76, 77(n2)
Brett, D. 158
British Tourist Authority 17
Britton, R. A. 27
Britton, S. G. 102, 103, 147
brochures 24–5, 27–8, 114–16
Bromley, R. 68
Brown, G. H. 142
Budihardjo, E. 54
Burt, E. 123
Butler, R. 155, 159
Butler, R. W. 123, 127, 129, 137

Campbell, C. 103
Canada: Arctic 105–11; Government of
  Northwest Territories 101, 108–9;
  Internet links 112–14; Inuit
  communities 101, 105–6, 111,
  112–16; Rocky Mountains 80–1,
  86–91

"Canada's North Now: The Great Betrayal" (Mowat) 106–7
Canadian Pacific Railway 86–7, 89, 90–1
capitalism: commodities 36; dominance 150; global 140–5; in Pacific 141; production 37–8
Cavell, E. 88
center/periphery dichotomy 147, 150
Classicism 141
Clifford, J. 148, 149
codes of conduct for tourists 115
commodification: Benjamin 37; cultural landscapes 102; escape 42–4; exploitation 37; fantasy 103; modernity 149; natural environment 9, 103; space 35
commodities: accumulation 38–9; as exports 146; Marx 36
communications network 40
communities: construction of place 112–16; ex-industrial 43, 44; ex-primitive 43, 44; power structures 117; pressures on 46; virtual 33–4
community-based tourism 108, 114
consciousness: collective 38; false 37
conservation 42–3
consumption 23; accumulation 38–9; commodification of landscape 102; seductive 36–7; tourist experience 104; virtual hyper-consumers 23–4
Conzen, M. 7
Cook, Thomas 34
Cook's tours, Scotland 127, 129–30, 131
Coppock, J. T. 18
counter-cultures 45
Craik, J. 4, 35
Crick, M. 18, 146
Cukier, J. 59, 61, 66, 71
Cukier-Snow, J. 63, 66
cultural geography 3
cultural heritage 1, 134, 135
cultural imperialism 147, 148
cultural landscapes 5, 7, 102, 140, 160
cultural politics 148
cultural production 83
cultural studies 46, 103–4
cultural tourism 1–2, 52–3, 67, 75–6
culture: authenticity 4, 8; Bali 53–6, 73, 74–5; employment in tourism 63; hegemonic 160; homogenization 149–50; and identity 9; Inuit

communities 111, 112; Sami people 156–8, 168–9; Victorian 34
Curr, E. 144
cyberspace 33, 34–5, 40–1
cybertourism 40–1
cybertravel 34
cyborgs 33

Dartington Amenity Research Trust 29
data analysis 85–6
Dead Seal (Purdy) 107
Defoe, Daniel 124
destination communities 2–3, 4
destinations: alternative 46; discourses 82–6; factors for success 121; global capitalist system 140–1; local people 7; meanings 84; production of space 155–60; Scotland as 121–2, 127–35; socially constructed 5–8, 82–3, 92, 115; transformed 154, 155, 158–60, 163
Dialla, A. 112
Dibnah, S. 56, 57
Dilley, R. S. 27
discourse analysis 85
discourses of destinations 82–6
Douglas, Ngaire 144, 145
Douglas, Norman 144, 145
dreams 19

Ebery, M. G. 66
Edinburgh 133
Elkan, W. 64
employment in tourism: migrants 71; socio-cultural implications 63, 74–5; status 64–6, 67–8
English Tourist Board 28–9
escape experience: commodified 42–4; virtual reality 41
ethnic identity 1
ethnic tourism 111, 169
evangelism, and tourism 17–18
ex-industrial community 43, 44
ex-primitive community 43, 44
excluded people 40–1
excursion routes 57
exoticism 141
exploitation 37, 106
exploration 141

Fairclough, N. 85

false consciousness 37
fantasy 103
Farrell, B. 145
Feifer, M. 45
Feuerstein, M. T. 65
Fiji 144
Fiji Publicity Board 144
Finlay, M. 45
Finney, C. M. 143
Finnish Lapland 154, 156; see also
    Saariselkä region
fjeld landscapes 162
Fladmark, M. 30
Flaherty, R. 105
foodstuffs, tourism 57
Forbes, D. K. 66
Fowler, P. J. 21
Francillon, G. 74–5
Friedman, M. 112
Friends of Children association, Lapland
    165–6

Gale, F. 6
Geertz, C. 67
Geertz, H. 67
gender: space 8; and tourism 65, 84, 86,
    87–91; see also women
geography 18–20; cultural 3; human
    6–7, 83; social and cultural 83; and
    tourism studies 81–2, 83
Geriya, I. W. 74
Gerry, C. 68
Gertler, L. 68
Giddens, A. 148
Gissing, George 35
Glasgow 133
global capitalist system 140–5
globalization: localization 150; Pacific
    145–51
Gold, J. R. 122, 125, 135, 136
Gold, M. M. 122, 125, 135, 136
golf 137
Goss, J. D. 27
Gould, Glenn 106
Greenwald, B. 105–6
Greenwood, D. 116
Grekin, J. 109–10, 112
Grise Fiord Hamlet 115
guidebooks 24–6; Lonely Planet 45–6
guides, Bali 68–73; case study 70–3;
    demographics 71; job satisfaction
    71–2; life-histories 72–3; status 63, 65

Hall, C. M. 103, 143, 149, 150
Hamilton, R. W. 109
Haraway, Donna 33
Harding, Alice 89
Harrison, D. 76
Harvey, D. 155
Haussmann, Baron 36
Hawaiian Islands 27, 145
Haywood, K. M. 155
hegemony, cultural 160
heritage industry: cultural 1; ex-
    industrial community 43–4;
    idealization of past 149; nostalgia
    21, 44; presentations 21
Hewison, R. 44
Highland Clearances 123
Hinman, Carolyn 88
history, tourism 82, 86–7
Hollinshead, K. 108, 112, 116
home/abroad concept 34
homogenization: cultural 149–50; of
    tourism 42
Hoogvelt, A. 141
Hudson, R. 64
Hughes, G. 9, 83
Hughes-Freeland, F. 148
Hull, J. 114
Hull, R. B. 51
human geographies 6–7, 83
Hunter, C. J. 6
Huysmans, J. K. 39
hyper-consumers 23–4
hyper-reality 21–2

idea of north 102, 105–7, 116
"The Idea of North" (Gould) 106
identity: culture 9; ethnic 1; modernity
    36, 148–9; Pacific 144–5; spatial
    19–20
illustrations, brochures 24–5
image: advertisements 102–3;
    consumption 23; media 105, 137;
    new technology 104; paradise 56,
    140; regions 28; Scotland 126–7;
    tourism 10, 104
imperialism 146–7
India, planning regulations 42, 47(n6)
indigenous peoples 168, 169; see also
    Inuit community; Sami people
Indonesia 52–3; social stratification 65;
    see also Bali
informal sector, tourism 65–6, 70, 73, 75

information processing models 26
infrastructure of tourism 44–5
institutionalization of regions 154,
    158–60
interior design 39
Internet: access to 113–14; cyberculture
    33; Inuit culture 112–14; multi-user
    domains 41; tourist images 104
Inuit communities: culture 111, 112;
    Internet 112–14; media images
    105–6; power structures 117;
    problems 106; tourism industry 112,
    114–16; visitor numbers 101;
    wastage 111

Jackson, P. 102, 104
Jackson, W. 7, 10
Jacobs, P. 117
Jacques, M. 150, 151
Java South, gender and tourism 65
Jencks, C. 39
Johnson, Samuel 124, 126
Johnston, M. 116
Johnston, R. J. 9
Jokinen, E. 84
journals, visitors 123

Kaltenborn, B. 107
Kaunispää 162, 165
Kemppinen, Kullervo 162–3, 164
Knudson, D. M. 52
Kulchyski, P. 112
Kuta, guides 70–3

land use 18, 167
landscape: Bali 52, 53, 66; commodified
    102; communities/tourists 102;
    constructed 2, 7–8; cultural 5, 7,
    102, 140, 160; meaning 55; and
    place 5; preservation/change 51;
    regional 53; stereotyped 158
Landseer, Sir Edwin Henry 127
Lapland 154, 156, 160–9
Lash, S. 8, 19
Laurier, E. 140
Le Corbusier 39
Lefebvre, H. 19, 34, 35–6, 39, 42, 156
Legault, S. 80
Lehr, J. C. 92
leisure, shopping 36–7
leisure imperialism 146

Lew, A. A. 30
Lichtheim, G. 147
liminality 22
literary visitors, Scotland 123–7
local people: destinations 7; interaction
    with tourists 51–2, 112; landscape
    102; values 156
localization 150
Lonely Planet guidebooks 45–6
Long, V. 58, 59
Loos, Adolf 39
Lopez, Barry 106–7
Lovel, H. 65
ludic space 19, 21, 24
Lumsdon, L. 17

MacArthur, Mrs Elizabeth 142
MacCannell, D. 21, 35, 42, 43, 44, 84,
    102, 111, 160, 163
McDonaldization 47(n7)
McGuigan, J. 33–4
machine metaphor 40, 47(n4)
McPherson, James 124–5
Magnus, Olaus 156
Marcuse, H. 42
marginalization 169–70
marketing, distortions 27
marketing of place 19
Marshall, A. J. 143
Marx, Karl 36;
    commodification/exploitation 37
Marxism, globalization 141
mass tourism 42, 52–3, 130
Maundeville, Sir John 24, 35, 46(n2)
meanings and values 20–4, 55, 84
media, and images 105–6, 137
Mendelssohn, Felix 125
Methodism 17
migrant workers, tourism 71
migration, and tourism 144
Milne, S. 109–10, 112, 113, 114, 115,
    145, 150
modernity: commodification 149;
    identity 36, 148–9; machine image
    40; tourism 148
modernization, Bali 61
Moss, J. 105, 107
Mowat, Farley 106–7
Mowl, G. 8
multi user domains 41
Murphy, P. E. 76
Murray, John 24–5

myths 19, 28

Nanook of the North (Flaherty) 105
Nash, D. 51, 146–7
national symbols, Bali 60
nature, tourism's use of 9, 103
Nature Canada 111
neo-colonialism 51
New Zealand, Tolaga Bay 142
noble savage image 105, 141–2
north, idea of 102, 105–7, 116
North of England Open Air Museum,
    Beamish 43–4
nostalgia, heritage 21, 44
Nunavut Territory 102, 108, 114
NWT Virtual Explorers' Guide 113

Ogilvy, D. 27
oil industry, Scotland 133, 134
Orr, W. 125
Ortner, S. B. 148
Ossian 124–5
otherness 103, 140
Ousby, I. 24
outdoor activity holidays, Scotland
    134–5, 136

Paasi, A. 154, 158–9
Paasilinna, Erno 169
Pacific: commodity exports 146;
    globalization 145–51; identity
    144–5; natural resources 146;
    paradise image 140; social
    construction 141
Page, S. 149
Palmer, E. 80
paradise image 56, 140
Paris, Arcades project 36–7
Paris, Ivy 89
Parkinson, Sydney 142
Partanen, S. J. 165, 167
Pausanias 24
Pemble, J. 34, 35
performativity 25, 43
Petsamo area, Finland 162, 165
phantasmagoria 36, 37, 38
Picard, M. 60–1, 67, 70
Pizam, A. 65
place: aesthetic management 21–2; and
    community 112–16; image 102–3;

landscape 5; representation 19–20;
    and space 19
planning regulations 42; Bali 59–60;
    India 42, 47(n6)
Plant, S. 33
pollution, Bali 58–9
Polo, Marco 24, 35
Pongsapich, A. 65
pony-trekking 135
Porter, M. E. 150
post-tourism 45
Poster, M. 33, 40, 41
postmodernity, tourist gaze 111
Powell, J.M. 143
power structures 117
Pratiwi, W. 65
Prebble, J. 123
primitive life, tourism 43
private/public space 39, 40
production systems 37–8, 83
promotional literature: advertisement
    26–7; Arctic 107–8; brochures 24–5,
    27–8, 114–16; Canadian
    government 108–9, 110; community
    initiatives 114–16; guidebooks
    24–6; representation of places
    19–20; Vanuatu 145
public/private space 39, 40
Purdy, Al 107

race, tourism 84
reality: paramount 41–2; tourism 154,
    158; virtual 40–1
reconstructions 43
Reed, Kate 89
regions: geographical concept 155;
    images 28; institutionalized 154,
    158–60; landscape 53
Regnard, Jean-François 156
Reimer, G. 112
religion, Bali 53, 60–1, 66
representation: authenticity 158; local
    168–9; of place 19–20; of space
    155–8; of women 90–1, 156–8, 166
reproduction 22
Revell, G. R. B. 51
Rheingold, H. 33, 40
Riley, C. 28
Ritzer, G. 47(n7)
Robins, K. 33
Rocky Mountains of Canada 80–1;
    gender and tourism history 86–7; as

honeymoon destination 91;
representations of women 90–1;
social constructions 80–1; women as
travelers 87–9; women as workers
89–90
Ross, Major Robert 143
Rossiter, J. 26, 27
Rousseau, Jean-Jacques 141
Royal Deeside 131

Saarinen, J. 155, 160
Saariselkä region 154; gold-mining
161–2; historical background 161–5;
land-use 167; local representations
168–9; marginalization 169–70;
tourism development 165–7; tourist
numbers 160, 163, 165, 167; Urho
Kekkonen National Park 167
Said, Edward 170
Saisselin, R. 37
Salmond, J. E. 124
Sami people 154, 156; culture 156–8,
168–9; protests 168
Samuel, R. 44
Sanur, guides 70–3
sawah (rice fields) 53, 54, 58
Schäffer Waren, Mary 87–8
Schama, S. 102
Schuett, M. A. 28
Schutz, A. 41–2
Scotland 123–7; access 126, 129,
130–4; accommodation 126, 129,
134; automobile age 132–4; Cook's
tours 127, 129–30, 131; cultural
heritage 134, 135; financing for
tourism 133, 134; golf 137;
hunting/shooting parties 125; image
126–7; infrastructure lacking 129;
lowlanders/highlanders 125; mass
tourism development 130; media
representations 137; as national
entity 136; natural sites of interest
128; outdoor activity holidays
134–5, 136; pre-steam era 128–9;
skiing holidays 134; steam age
129–32; stereotypes 25, 126, 136; as
tourist destination 121–2, 127–35;
visitors 123–7; walking holidays
131–2; whisky 135, 136
Scotland, places: Balmoral 127, 131;
Edinburgh 133, 135; Glasgow 133,

135; Royal Deeside 131; Spey Valley
131, 135; west coast 130
Scott, Sir Walter 125, 126–7
Scottish Tourist Board 29, 136
Scottish Youth Hostels Association 132
self-employment in tourism 68
Selwyn, T. 28
Shadow of the Wolf (Dorfman) 106
Shefferus, Johan 156
Shields, R. 22, 155–6
shop window study 47(n5)
shopping experience 36–7
Shortridge, J. R. 19
sight sacralization theory 160
sign, economy of 23
Silbergh, D. 29–30
skiing holidays 134
Smith, B. 142
Smith, C. 88
Smith, V. 116
social consequences, tourism 1, 36
social construction: destinations 5–8,
82–3, 92, 155; idea of north 116;
landscape 5, 7; Pacific 141; and
perception 4; Rocky Mountains 80–1
social and cultural geography 83
social perceptions 3, 4
social status 67–8
space: commodified 35; destinations
155–60; differentiated 20, 30;
gendered 8; human constructions 6;
hyper-reality 21–2; identity 19–20;
ludic 19, 21, 24; and place 19;
representations 155–8; tourist
pressures 46
Spackman, N. 28
spatialization 24, 155
Spey Valley 131, 135
Squire, S. 102, 104
Starch Inra Hooper Inc. 26
steamships 129–30
stereotyping: landscape 158; Scotland
25, 126, 136
Strauss, A. L. 85
sustainability 82
symbols, national 60

Tahiti 142
tartan image 126, 136
Telford, Thomas 126
Tench, Captain Watkin 142
Thailand, guides 65

themed trails 29
themes, tourism 21, 29
Thomas, R. 41
Thomson Holidays' brochures 28
Tocher, M. 80
Tolaga Bay 142
Tomlinson, J. 147, 148, 149, 150
topos 27, 165
tourism: billiard ball model 1, 5; cultural
    politics 148; economic importance 1,
    52–3; employment 63, 64–6, 67–8;
    and evangelism 17–18; gender factors
    65, 84, 86, 87–91; geography of
    18–20; ideology 102; images 10, 104;
    informal sector 65–6, 70, 73, 75;
    infrastructure 44–5, 129; interaction
    of tourists/locals 1, 36, 43, 57, 88,
    112; landscape construction 2, 7–8; as
    leisure imperialism 146; life-cycle
    155; meanings and values 20–4, 84;
    and migration 144; modernity 148;
    phantasmagoria 36, 37, 38; physical
    movement 34; preservation/change
    51; reality 154, 158; themes 21, 29;
    and travel 128; women workers 89–90
tourism, types: community-based 108,
    114; cultural 1–2, 52–3, 67, 75–6;
    cybertourism 40–1; ethnic 111, 169;
    heritage 149; mass 42, 52–3, 130;
    whisky-based 135, 136
tourism development: evolution cycle
    159; Saariselkä 165–7;
    spatial/economic 85; women 165, 166
tourism geography 83
tourism history 82, 86
tourism studies 81–2, 93
tourist flows 35, 44–5
tourist surveys 109–10
tourist trails 29–30
tourists: Arctic perceptions 109–11;
    behavior, acceptability 22; codes of
    conduct 115; interaction with locals
    51–2, 112; *see also* visitors
Towner, J. 8
Townsend, A. 64
training in tourism, Bali 75–6, 77(n2)
transformation of destination 154, 155,
    158–60, 163
travel experience 102, 128
Tuan, Yi-Fu 3
Turner, Joseph 127

Turner, L. 64
Turner, V. 22

UNDP 66
UNESCO 42, 64–5
Urho Kekkonen National Park 167
Urry, J. 8, 19, 34, 45, 64, 83, 103, 104,
    155, 158

values 20–4, 55, 84, 156
Vanuatu 145
Vaux, Mary 88
Veijola, S. 84
Vickers, A. 70
Victoria, Queen 127
Victorian culture 34
Vidler, A. 40
virtual communities 33–4
virtual hyper-consumers 23–4
virtual reality 40–1
visitors: behavioral guidelines 109;
    literary 123–7; and locals 51–2; *see
    also* tourists

wage work, tourism 68
Wagstaff, J. M. 7
Wall, G. 52, 56, 57, 58, 59, 60, 63, 66,
    71
Wallerstein's World Systems Theory 141
wastage, Inuit communities 111
Wenzel, G. 111, 116
Wesley Trail 17
whisky-based tourism 135, 136
Wicks, B. E. 28
Wijaya, Ngurah 69
wilderness 123, 163, 165
Wilkinson, P. F. 65
Williams, R. 37, 160
Wilson, R. 33
women: representations 90–1, 156–8,
    166; and tourism development 165,
    166; travelers/explorers 87–9;
    workers 89–90
Wood 47(n5)
Wood, R. E. 148
wood carving, Bali 58
Wordsworth, William 24
Wright, P. 44

Youngson, A. J. 123